Principles and Practices in Spice Crops Cultivation

Principles and Practices in Spice Crops Cultivation

R.C. Upadhyaya

ANMOL PUBLICATIONS PVT. LTD.
NEW DELHI-110 002 (INDIA)

ANMOL PUBLICATIONS PVT. LTD.
H.O.: 4374/4B, Ansari Road, Darya Ganj,
New Delhi-110 002 (India)
Ph.: 23278000, 23261597
B.O.: No. 1015, Ist Main Road, BSK IIIrd Stage
IIIrd Phase, IIIrd Block
Bangalore - 560 085 (India)
Visit us at: www.anmolpublications.com

Principles and Practices in Spice Crops Cultivation

First Published, 2008

ISBN 978-81-261-3425-0

PRINTED IN INDIA

Printed at Mehra Offset Press, Delhi.

Contents

Preface vii

1. **Species** **1**
Tree Species • American Chestnut • American Hornbeam • Basswood • Big Toothed Aspen • Black Birch • Black Cherry • Black Locust • Black Maple • Black Oak • Black Walnut • Black Willow • Butternut • Chestnut Oak • Crabapple • Cucumbertree • Eastern Hemlock • Flowering Dogwood • Hawthorn • Mockernut Hickory • Pitch Pine • Redcedar • Red Maple • Red Oak • Red Pine • Red Spruce • Sassafras • Serviceberry • Shagbark Hickory • Striped Maple • Sugar Maple • Tamarack • Tulip Tree • White Ash • White Oak • White Pine • Yellow Birch • Full Listings

2. **Precautionary Approach to Species** **73**
Research and Technology • Management • Spices and Condiments • Extinction Event

3. **Management of Invasive Alien Species** **107**
Information Digest of Section • Convention and Invasive Alien Species • Relation to Invasive Alien Species • Cooperation between International Bodies • Future Activities

4. **Herbal Sources of Tibetan Medicines** **135**
New Dimensions In Cultivating Medicinal Plants • Origin of Biodynamic Agriculture • Principles of Biodynamic Cultivation • Polarities in Biodynamics • Earth and Sun • Biodynamic

Preparations • Geometric Principles and Medicinal Values • Cosmic Integration • Soil Fertility of Plant Nutrition

5. **New Dinosaur Species Found in India** **157**
Continental Drift • Spices and Condiments—Nutmeg • Coriander • Garlic • Curry Leaf • Tamarind • Allspice • Vanilla • Fenugreek • Clove • Cardamom • Introduced Species • Invasive Species • Superior Competition • Facilitation • Ecosystem based Mechanisms • Ecology of Invasive Species • Traits of Invasive Species • Traits of Invaded Ecosystems • Vectors • Impacts of Invasive Species • Economic Impacts • Agriculture • Forestry • Tourism and Recreation • Health Impacts • Mechanical Control • Chemical Control • Biological Control • Prevention • Seed

Bibliography 231

Index 237

Preface

Horticulture has a special significance for the economic development. Most of the people, though hard working and face the calamities of nature and live in harsh conditions, yet produce lot of fruits, vegetables and other flowering and ornamental plants including the medicinal and aromatic plants. In order to increase the productivity of horticultural crops growers must take the advantage of the expertise of scientists. Keeping this idea in focus, the present work has been compiled which contains experiences of the experts and related people. The volumes cover advances in strategies, production, plant protection, value addition and other important areas of horticulture development.

The Encyclopaedia provides practical information on production technologies of tropical and subtropical fruits as well as plantation crops and horticultural nursery. Post-harvest management, marketing and export trends have been furnished for guidance. It also presents most realistic cost estimates and returns from cultivation of such crops. The volumes aim at viewing commercial prospects for horticulture. It has been carefully designed to meet the queries of intending investors, financial institutes, creditors, traders and growers engaged in dealing with horticultural crops. Further it will serve the purpose reference books to students, academicians and related people.

R.C. Upadhyaya

1

Species

In biology, a species is one of the basic units of biological classification. A usable definition of the word "species" and reliable methods of identifying particular species are essential for stating and testing biological theories and for measuring bio-diversity. A species consists of individual organisms which are very similar in appearance, anatomy, physiology and genetics due to having relatively recent common ancestors. Traditionally, multiple examples of a proposed species must be studied for unifying characters before it can be regarded as a species.

The commonly used names for types of animal in English and many other languages often do not correspond to species. For example: dogs are a sub-species of wolf; but the word "deer" refers to a family of 34 species, such as the Red Deer and Mule Deer.

Binomial Convention for Naming Species

In scientific classification, a species is assigned a two-part name, treated as Latin. The genus is listed first (with its leading letter capitalized), followed by a second term. For example, gray wolves belong to the species *Canis lupus*, coyotes to *Canis latrans*, golden jackals to *Canis aureus*, etc.; and all of these belong to the genus *Canis* (which also contains many other species). The name of the species is the whole binomial, not just the second term (which may be called specific name for animals).

Books and articles sometimes intentionally do not identify species fully and use the abbreviation "*sp.*" in the singular or "*spp.*" in the plural in the place of the specific epithet, for example *Canis sp.* This commonly occurs in the following types of situation:

- The authors are confident that some individuals belong to a particular genus but are not sure to which exact species they belong. This is particularly common in paleontology.
- The authors use the "*spp.*" as a short way of saying that something applies to many species within a genus, but do not wish say that it applies to all species within that genus. If scientists mean that something applies to all species with a genus, they use the genus name without the specific epithet.

In books and articles which use the Latin alphabet, genus and species names are usually printed in italics.

The binomial naming convention, later formalized in the biological codes of nomenclature, was first used by Leonhart Fuchs and introduced as the standard by Carolus Linnaeus in the 1700s and as a result is sometimes called the "binomial nomenclature". At this time, the chief biological theory was that species represented independent acts of creation by God, and were therefore considered objectively real and immutable.

Difficulty of Defining "species" and Identifying Particular Species

It is surprisingly difficult to define the word "species" in a way that applies to all naturally occurring organisms, and the debate among biologists about how to define "species" and how to identify actual species is called the species problem.

Most textbooks define a species as all the individual organisms of a natural population that generally interbreed at maturity in the wild and whose interbreeding produces fertile offspring. Various parts of this definition are there to exclude some unusual or artificial matings:

- Those which occur only in captivity (when the animal's normal mating partners may not be available) or as a result of deliberate human action.
- Animals which may be physically and physiologically capable of mating but do not normally do so because only

their normal mating partners perform the courtship rituals or some other behaviour "correctly".

- Animals whose offspring are normally sterile. For example, mules and hinnies have never (so far) produced further offspring when mated with a creature of the same type (a mule with a mule, or a hinny with a hinny).

Difficulties in Defining "Species" for Living Organisms

The typical textbook definition (above) works well for most multi-celled organisms, but there are several types of situations where it breaks down:

- By definition it applies only to organisms which reproduce sexually. So it does not work for asexually reproducing single-celled organisms and for the relatively few parthenogenetic multi-celled organisms.
- Some hybrids, *e.g.* mules, hinnies, ligers and tigons, apparently cannot produce offspring when mated with one of their own kind (*e.g.* a mule with a mule), but sometimes do produce offspring when mated with members of one of the parent species (*e.g.* a ligon with a lion). Usually in such hybrids the males are sterile, so one could improve the basic textbook definition by changing "... whose interbreeding produces fertile offspring" to "... whose interbreeding produces offspring in which both sexes are normally fertile".
- In ring species (also known as clines), members of adjacent populations interbreed successfully but members of widely-separated populations do not.
- In a few cases it may be physically impossible for animals which are members of the same species to mate, for example a Great Dane and a Chihuahua are both dogs and therefore members of the same species (which also includes gray wolves), but cannot mate because of the great difference in size.

Horizontal gene transfer makes it even more difficult to define the word "species". There is strong evidence of horizontal gene transfer between very dissimilar groups of procaryotes, and possibly between dissimilar groups of single-celled eucaryotes; and Williamson argues that there is evidence for it in some crustaceans and echinoderms. All definitions of the word "species" assume that an organism gets all its genes from one or two parents which are very like that organism, but horizontal gene transfer makes that assumption false.

Difficulties in Defining "Species" for Extinct Organisms

Many extinct organisms are known only from fossils, which generally only preserve features. Fossils have not (so far) shown us what bred with what, and cannot tell us whether any resulting offspring would have been fertile. So paleontologists generally use either the morphological or the cladistic definition of species.

Paleontologists also have to cope with another difficulty: one species may gradually evolve into one or more others after a few million years; the original type of organism and the final one are so different that one could not regard the ancestors and the descendants as members of the same species if they existed at the same time; but the intermediate types are so similar to the next and previous types that one cannot say exactly where species A changed into species B. Paleontologists devised the concept of chronospecies to describe the simplest case, where at the end of the process there is only one descendant type of organism and there are no longer any individuals of the ancestral type. But even this refinement does not work in cases where several descendant types are alive at the same time or where the ancestral type and at least one descendant type are alive at the same time - and both of these situations are common in the evolution of life on Earth. Human evolution may offer a striking example: some geneticists have suggested that for about 1 million years there was some interbreeding between the early ancestors of humans and the early ancestors of chimpanzees.

Numbers of Species

As a soft guide, however, the numbers of identified modern species as of 2004 can be broken down as follows:

- 287,655 plants, including:
  - o 15,000 mosses,
  - o 13,025 ferns,

  - o 980 gymnosperms,
  - o 199,350 dicotyledons,
  - o 59,300 monocotyledons;
- 74,000-120,000 fungi;
- 10,000 lichens;
- 1,250,000 animals, including:
  - o 1,190,200 invertebrates:
    - — 950,000 insects,
    - — 70,000 mollusks,
    - — 40,000 crustaceans,
    - — 130,200 others;
  - o 58,808 vertebrates:
    - — 29,300 fish,
    - — 5,743 amphibians,
    - — 8,240 reptiles,
    - — 10,234 birds, (9799 extant as of 2006)
    - — 5,416 mammals.

However the *total* number of species for some phyla may be much higher:

- 5-10 million bacteria;
- 1.5 million fungi;

Definitions of Species

The question of how best to define "species" is one that has occupied biologists for centuries, and the debate itself has become known as the species problem. One definition that is widely used is that a species is a group of actually or potentially interbreeding populations that are reproductively isolated from other such groups.

The definition of a species given above is due to Mayr, and is somewhat idealistic. Since it assumes sexual reproduction, it leaves the term undefined for a large class of organisms that reproduce asexually. Biologists frequently do not know whether two morphologically similar groups of organisms are "potentially" capable of interbreeding. Further, there is considerable variation in the degree to which hybridization may succeed under natural and experimental conditions, or even in the degree to which some organisms use sexual reproduction between individuals to breed. Consequently, several lines of thought in the definition of *species* exist:

Typological Species

A group of organisms in which individuals are members of the species if they sufficiently conform to certain fixed properties. The clusters of variations or phenotypes within specimens (*i.e.* longer and shorter tails) would differentiate the species. This method was used as a "classical" method of determining species, such as with Linnaeus early in evolutionary theory. However, we now know that different phenotypes do not always constitute different species (*e.g.*: a 4-winged Drosophila born to a 2-winged mother is not a different species). Species named in this manner are called *morphospecies*.

Morphological Species

A population or group of populations that differs morphologically from other populations. For example, we can distinguish between a chicken and a duck because they have different shaped bills and the duck has webbed feet. Species have been defined in this way since well before the beginning of recorded history. This species concept is much criticised because more recent genetic data reveal that genetically distinct populations may look very similar and, contrarily, large morphological differences sometimes exist between very closely-related populations. Nonetheless, most species known have been described solely from morphology.

Biological/Isolation Species

A set of actually or potentially interbreeding populations. This is generally a useful formulation for scientists working with living examples of the higher taxa like mammals, fish, and birds, but meaningless for organisms that do not reproduce sexually. It does not distinguish between the theoretical possibility of interbreeding and the actual likelihood of gene flow between populations and is thus impractical in instances of allopatric (geographically isolated) populations. The results of breeding experiments done in artificial conditions may or may not reflect what would happen if the same organisms encountered each other in the wild, making it difficult to gauge whether or not the results of such experiments are meaningful in reference to natural populations.

Biological/reproductive Species

Two organisms that are able to reproduce naturally to produce fertile offspring. Organisms that can reproduce to almost always make infertile hybrids, such as a mule or hinny, are not considered to be the same species.

Mate-recognition Species

A group of organisms that are known to recognize one another as potential mates. Like the isolation species concept above, it applies only to organisms that reproduce sexually. Unlike the isolation species concept, it focuses specifically on pre-mating reproductive isolation.

Phylogenetic (Cladistic)/Evolutionary/Darwinian Species

A group of organisms that shares an ancestor; a lineage that maintains its integrity with respect to other lineages through both time and space. At some point in the progress of such a group, members may diverge from one another. When such a divergence becomes sufficiently clear, the two populations are regarded as separate species. Subspecies as such are not recognized under this approach; either a population is a phylogenetic species or it is not taxonomically distinguishable.

Ecological Species

A set of organisms adapted to a particular set of resources, called a niche, in the environment. According to this concept, populations form the discrete phenetic clusters that we recognize as species because the ecological and evolutionary processes controlling how resources are divided up tend to produce those clusters.

Genetic Species

based on similarity of DNA of individuals or populations. Techniques to compare similarity of DNA include DNA-DNA hybridization, and genetic fingerprinting (or DNA barcoding).

Phenetic Species

Based on phenotypes.

Recognition Species

Based on behavioral interactions.

Microspecies

Species that reproduce without meiosis or fertilization so that each generation is genetically identical to the previous generation.

Cohesion Species

Most inclusive population of individuals having the potential for phenotypic cohesion through intrinsic cohesion mechanisms. This is an expansion of the mate-recognition species concept to allow for post-mating isolation mechanisms; no matter whether populations can hybridize successfully, they are still distinct cohesion species if the amount of hybridization is insufficient to completely mix their respective gene pools.

Evolutionary Significant Unit (ESU)

An evolutionary significant unit is a population of organisms that is considered distinct for purposes of conservation. Often referred to as a species or a *wildlife species*, an ESU also has several possible definitions, which coincide with definitions of species.

In practice, these definitions often coincide, and the differences between them are more a matter of emphasis than of outright contradiction. Nevertheless, no species concept yet proposed is entirely objective, or can be applied in all cases without resorting to judgement. Given the complexity of life, some have argued that such an objective definition is in all likelihood impossible, and biologists should settle for the most practical definition. For most vertebrates, this is the biological species concept (BSC), and to a lesser extent (or for different purposes) the phylogenetic species concept (PSC). Many BSC subspecies are considered species under the PSC; the difference between the BSC and the PSC can be summed up insofar as that the BSC defines a species as a consequence of manifest evolutionary *history*, while the PSC defines a species as a consequence of manifest evolutionary *potential*. Thus, a PSC species is "made" as soon as an evolutionary lineage has started to separate, while a BSC species starts to exist only when the lineage separation is complete.

Importance in Biological Classification

The idea of *species* has a long history. It is one of the most important levels of classification, for several reasons:

- It often corresponds to what lay people treat as the different basic kinds of organism - dogs are one species, cats another.
- It is the standard binomial nomenclature (or trinomial nomenclature) by which scientists typically refer to organisms.
- It is the highest taxonomic level which mostly cannot be made more or less inclusionary.

After thousands of years of use, the concept remains central to biology and a host of related fields, and yet also remains at times ill-defined.

Implications of Assignment of Species Status

The naming of a particular species should be regarded as a *hypothesis* about the evolutionary relationships and distinguishability of that group of organisms. As further information comes to hand, the hypothesis may be confirmed or refuted. Sometimes, especially in the past when communication was more difficult, taxonomists working in isolation have given two distinct names to individual organisms later identified as the same species. When two named species are discovered to be of the same species, the older species name is usually retained, and the newer species name dropped, a process called *synonymization*, or convivially, as lumping.

Dividing a taxon into multiple, often new, taxons is called splitting. Taxonomists are often referred to as "lumpers" or "splitters" by their colleagues, depending on their personal approach to recognizing differences or commonalities between organisms.

Traditionally, researchers relied on observations of anatomical differences, and on observations of whether different populations were able to interbreed successfully, to distinguish species; both anatomy and breeding behaviour are still important to assigning species status. As a result of the revolutionary (and still ongoing) advance in microbiological research techniques, including DNA analysis, in the last few decades, a great deal of additional knowledge about the differences and similarities between species has become available. Many populations which were formerly regarded as separate species are now considered to be a single taxon, and many formerly grouped populations have been split. Any taxonomic level (species, genus, family, etc.) can be synonymized or split, and at higher taxonomic levels, these revisions have been still more profound.

From a taxonomical point of view, groups within a species can be defined as being of a taxon hierarchically lower than a species. In zoology only the subspecies is used, while in botany the variety, subvariety, and form are used as well. In conservation biology, the concept of evolutionary significant units (ESU) is used, which may be define either species or smaller distinct population segments.

In general, for large, complex, organisms that reproduce sexually (such as mammals and birds), one of several variations on the *isolation* or *biological* species concept is employed. Often, the distinction between different species, even quite closely related ones, is simple. Horses (*Equus caballus*) and donkeys (*Equus asinus*) are easily told apart even without study or training, and yet are so closely related that they can interbreed after a fashion. Because the result, a mule or hinny, is not usually fertile, they are clearly separate species.

But many cases are more difficult to decide. This is where the isolation species concept diverges from the evolutionary species concept. Both agree that a species is a lineage that maintains its integrity over time, that is diagnosably different from other lineages (else we could not recognise it), is reproductively isolated (else the lineage would merge into others, given the chance to do so), and has a working intra-species recognition system (without which it could not continue).

In practice, both also agree that a species must have its own independent evolutionary history—otherwise the characteristics just mentioned would not apply. The species concepts differ in that the evolutionary species concept does not make predictions about the future of the population: it simply records that which is already known. In contrast, the isolation species concept refuses to assign the rank of species to populations that, in the best judgement of the researcher, would recombine with other populations if given the chance to do so.

The Isolation Question

There are, essentially, two questions to resolve. First, is the proposed species consistently and reliably distinguishable from other species? Secondly, is it likely to remain so in the future? To take the second question first, there are several broad geographic possibilities.

The proposed species are sympatric—they occupy the same habitat. Observation of many species over the years has failed to establish even a single instance of two diagnostically different populations that exist in sympatry and have then merged to form one united population. Without reproductive isolation, population differences cannot develop, and given reproductive isolation, gene flow between the populations cannot merge the differences. This is not to say that cross breeding does not take place at all, simply that it has become negligible. Generally, the hybrid individuals are less capable of successful breeding than pure-bred individuals of either species.

The proposed species are allopatric—they occupy different geographical areas. Obviously, it is not possible to observe reproductive isolation in allopatric groups directly. Often it is not possible to achieve certainty by experimental means either:

even if the two proposed species interbreed in captivity, this does not demonstrate that they would freely interbreed in the wild, nor does it always provide much information about the evolutionary fitness of hybrid individuals. A certain amount can be inferred from other experimental methods: for example, do the members of population A respond appropriately to playback of the recorded mating calls of population B? Sometimes, experiments *can* provide firm answers. For example, there are seven pairs of apparently almost identical marine snapping shrimp (*Alpheus*) populations on either side of the Isthmus of Panama, which did not exist until about 3 million years ago. Until then, it is assumed, they were members of the same seven species. But when males and females from opposite sides of the isthmus are placed together, they fight instead of mating. Even if the isthmus were to sink under the waves again, the populations would remain genetically isolated: therefore they are now different species. In many cases, however, neither observation nor experiment can produce certain answers, and the determination of species rank must be made on a 'best guess' basis from a general knowledge of other related organisms.

The proposed species are parapatric—they have breeding ranges that abut but do not overlap. This is fairly rare, particularly in temperate regions. The dividing line is often a sudden change in habitat (an ecotone) like the edge of a forest or the snow line on a mountain, but can sometimes be remarkably trivial. The parapatry itself indicates that the two populations occupy such similar ecological roles that they cannot coexist in the same area. Because they do not crossbreed, it is safe to assume that there is a mechanism, often behavioural, that is preventing gene flow between the populations, and that therefore they should be classified as separate species.

There is a hybrid zone where the two populations mix. Typically, the hybrid zone will include representatives of one or both of the 'pure' populations, plus first-generation and back-crossing hybrids. The strength of the barrier to genetic transmission between the two pure groups can be assessed by the width of the hybrid zone relative to the typical dispersal distance of the organisms in question. The dispersal distance of oaks, for example, is the distance that a bird or squirrel can be expected to carry an acorn; the dispersal distance of Numbats is about 15 kilometres, as this is as far as young Numbats will normally travel in search of vacant territory to occupy after leaving the nest. The narrower the hybrid zone relative to the dispersal distance, the less gene flow there is between the population groups, and the more likely it is that they will continue on separate evolutionary paths. Nevertheless, it can be very difficult to predict the future course of a hybrid zone; the decision to define the two hybridizing populations as either the same species or as separate species is difficult and potentially controversial.

The variation in the population is clinal; at either extreme of the population's geographic distribution, typical individuals are clearly different, but the transition between them is seamless and gradual. For example, the Koalas of northern Australia are clearly smaller and lighter in colour than those of the south, but there is no particular dividing line: the further south an individual Koala is found, the larger and darker it is likely to be; Koalas in intermediate regions are intermediate in weight and colour. In contrast, over the same geographic range, black-backed (northern) and white-backed (southern) Australian Magpies do not blend from one type to another: northern populations have black backs, southern populations white backs, and there is an extensive hybrid zone where both 'pure' types are common, as are crossbreeds. The variation in Koalas is clinal (a smooth transition from north to south, with populations in any given small area having a uniform appearance), but the variation in magpies is *not* clinal. In both cases, there is some uncertainty regarding correct classification, but the consensus view is that species rank is not justified in either. The gene flow between northern and southern magpie populations is judged to be sufficiently restricted to justify terming them subspecies (not full species); but the seamless way that local Koala populations blend one into another shows that there is substantial gene flow between north and south. As a result, experts tend to reject even subspecies rank in this case.

The Difference Question

Obviously, when defining a species, the geographic circumstances become meaningful only if the populations groups in question are clearly different: if they are not consistently and reliably distinguishable from one another, then we have no grounds for believing that they might be different species. The key question in this context, is "how different is different?" and the answer is usually "it all depends".

In theory, it would be possible to recognise even the tiniest of differences as sufficient to delineate a separate species, provided only that the difference is clear and consistent (and that other criteria are met). There is no universal rule to state the smallest allowable difference between two species, but in general, very trivial differences are ignored on the twin grounds of simple practicality, and genetic similarity: if two population groups are so close that the distinction between them rests on an obscure and microscopic difference in morphology, or a single base substitution in a DNA sequence, then a demonstration of restricted gene flow between the populations will probably be difficult in any case.

More typically, one or other of the following requirements must be met:

- It is possible to reliably measure a *quantitative* difference between the two groups that *does not overlap*. A population has, for example, thicker fur, rougher bark, longer ears, or larger seeds than another population, and although this characteristic may vary within each population, the two do not grade into one another, and given a reasonably large sample size, there is a definite discontinuity between them. Note that this applies to *populations*, not individual organisms, and that a small number of exceptional individuals within a population may 'break the rule' without invalidating it. The less a quantitative difference varies *within* a population and the more it varies *between* populations, the better the case for making a distinction. Nevertheless, borderline situations can only be resolved by making a 'best-guess' judgement.

- It is possible to distinguish a *qualitative* difference between the populations; a feature that does not vary continuously but is either entirely present or entirely absent. This might be a distinctively shaped seed pod, an extra primary feather, a particular courting behaviour, or a clearly different DNA sequence.

Sometimes it is not possible to isolate a single difference between species, and several factors must be taken in combination. This is often the case with plants in particular. In eucalypts, for example, *Corymbia ficifolia* cannot be reliably distinguished from its close relative *Corymbia calophylla* by any single measure (and sometimes individual trees cannot be definitely assigned to either species), but populations of *Corymbia* can be clearly told apart by comparing the colour of flowers, bark, and buds, number of flowers for a given size of tree, and the shape of the leaves and fruit.

When using a combination of characteristics to distinguish between populations, it is necessary to use a reasonably small number of factors (if more than a handful are needed, the genetic difference between the populations is likely to be insignificant and is unlikely to endure into the future), and to choose factors that are functionally independent (height and weight, for example, should usually be considered as one factor, not two).

Historical Development of the Species Concept

In the earliest works of science, a species was simply an individual organism that represented a group of similar or nearly identical organisms. No other relationships beyond that group were implied. Aristotle used the words *genus* and *species* to mean generic and specific categories. Aristotle and other pre-Darwinian scientists took the species to be distinct and unchanging, with an "essence", like the chemical elements. When early observers began to develop systems of organization for living things, they began to place formerly isolated species into a context. Many of these early delineation schemes would now be considered whimsical and these included consanguinity based on colour (all plants with yellow flowers) or behaviour (snakes, scorpions and certain biting ants).

In the 18th century Carolus Linnaeus classified organisms according to differences in the form of reproductive apparatus. Although his system of classification sorts organisms according to degrees of similarity, it made no claims about the relationship between similar species. At that time, it was still widely believed that there was no organic connection between species, no matter how similar they appeared. This approach also suggested a type of idealism: the notion that each species existed as an "ideal form". Although there are always differences (although sometimes minute) between individual organisms, Linnaeus considered such variation problematic. He strove to identify individual organisms that were exemplary of the species, and considered other non-exemplary organisms to be deviant and imperfect.

By the 19th century most naturalists understood that species could change form over time, and that the history of the planet provided enough time for major changes. Jean-Baptiste Lamarck, in his 1809 *Zoological Philosophy*, offered one of the first logical arguments against creationism. The new emphasis was on determining *how* a species could change over time. Lamarck suggested that an organism could pass on an acquired trait to its offspring, *i.e.*, the giraffe's long neck was attributed to generations of giraffes stretching to reach the leaves of higher treetops (this well-known and simplistic example, however, does not do justice to the breadth and subtlety of Lamarck's ideas). With the acceptance of the natural selection idea of Charles Darwin in the 1860s, however, Lamarck's view of goal-oriented evolution, also known as a teleological process, was eclipsed.

Recent interest in inheritance of acquired characteristics centers around epigenetic processes, *e.g.* methylation, that do not affect DNA sequences, but instead alter expression in an inheritable manner. Thus, neo-lamarckism, as it is sometimes termed, is not a challenge to the theory of evolution by natural selection.

Charles Darwin and Alfred Wallace provided what scientists now consider as the most powerful and compelling theory of evolution. Darwin argued that it was populations that evolved, not individuals. His argument relied on a radical shift in perspective from that of Linnaeus: rather than defining species in ideal terms (and searching for an ideal representative and rejecting deviations), Darwin considered variation among individuals to be natural. He further argued that variation, far from being problematic, actually provides the *explanation* for the existence of distinct species.

Darwin's work drew on Thomas Malthus' insight that the rate of growth of a biological population will always outpace the rate of growth of the resources in the environment, such as the food supply. As a result, Darwin argued, not all the members of a population will be able to survive and reproduce. Those that did will, on average, be the ones possessing variations—however slight—that make them slightly better adapted to the environment. If these variable traits are heritable, then the offspring of the survivors will also possess them. Thus, over many generations, adaptive variations will accumulate in the population, while counter-adaptive will be eliminated.

It should be emphasized that whether a variation is adaptive or non-adaptive depends on the environment: different environments favour different traits. Since the environment effectively selects which organisms live to reproduce, it is the environment (the "fight for existence") that selects the traits to be passed on. This is the theory of evolution by natural selection. In this model, the length of a giraffe's neck would be explained by positing that proto-giraffes with longer necks would have had a significant reproductive advantage to those with shorter necks. Over many generations, the entire population would be a species of long-necked animals.

In 1859, when Darwin published his theory of natural selection, the mechanism behind the inheritance of individual traits was unknown. Although Darwin made some speculations on how traits are inherited (pangenesis), his theory relies only on the fact that inheritable traits *exist*, and are variable (which makes his accomplishment even more remarkable.) Although Gregor Mendel's paper on genetics was published in 1866, its significance was not recognized. It was not until 1900 that his work was rediscovered by Hugo de Vries, Carl Correns and

Erich von Tschermak, who realised that the "inheritable traits" in Darwin's theory are genes.

The theory of the evolution of species through natural selection has two important implications for discussions of species—consequences that fundamentally challenge the assumptions behind Linnaeus' taxonomy. First, it suggests that species are not just similar, they may actually be related. Some students of Darwin argue that *all* species are descended from a common ancestor. Second, it supposes that "species" are not homogeneous, fixed, permanent things; members of a species are all different, and over time species change. This suggests that species do not have any clear boundaries but are rather momentary statistical effects of constantly changing gene-frequencies. One may still use Linnaeus' taxonomy to identify individual plants and animals, but one can no longer think of species as independent and immutable.

The rise of a new species from a parental line is called speciation. There is no clear line demarcating the ancestral species from the descendant species.

Although the current scientific understanding of species suggests that there is no rigorous and comprehensive way to distinguish between different species in *all* cases, biologists continue to seek concrete ways to operationalize the idea. One of the most popular biological definitions of species is in terms of reproductive isolation; if two creatures cannot reproduce to produce fertile offspring, then they are in different species. This definition captures a number of intuitive species boundaries, but it remains imperfect.

It has nothing to say about species that reproduce asexually, for example, and it is very difficult to apply to extinct species. Moreover, boundaries between species are often fuzzy: there are examples where members of one population can produce fertile offspring with a second population, and members of the second population can produce fertile offspring with members of a third population, but members of the first and third population cannot produces fertile offspring. Consequently, some people reject this definition of a species.

In recent years we have witnessed the drastic reduction in the size of breeding populations and the geographical range of many large mammals. In earlier times it was assumed that every species existed in at least a few thousand living individuals, except very rare relic, isolated groups. In the present, many well know mammal and bird species are so stressed by habitat loss, and other effects of the modern world, that only a very few breeding males may contribute the genetic material to a small number of breeding females. In these highly stressed conditions, the likelihood of change is very much greater. Mammals may become smaller, have darker fur, more stripes, more cautious behaviour, even over time learn to co-exist with the human world. Very likely, evolution is radically accelerated, and we are only beginning to notice it. It is possible that this severe stress is essential to the creation of new species, and may have been a prime factor throughout biological history, from other population reducing influences.

Richard Dawkins defines two organisms as conspecific if and only if they have the same number of chromosomes and, for each chromosome, both organisms have the same number of nucleotides (*The Blind Watchmaker*, p. 118). However, most if not all taxonomists would strongly disagree. For example, in many amphibians, most notably in New Zealand's *Leiopelma* frogs, the genome consists of "core" chromosomes which are mostly invariable and accessory chromosomes, of which exist a number of possible combinations. Even though the chromosome numbers are highly variable between populations, these can interbreed successfully and form a single evolutionary unit. In plants, polyploidy is extremely commonplace with few restrictions on interbreeding; as individuals with an odd number of chromosome sets are usually sterile, depending on the actual number of chromosome sets present, this results in the odd situation where some individuals of the same evolutionary unit can interbreed with certain others and some cannot, with all populations being eventually linked as to form a common gene pool.

The classification of species has been profoundly affected by technological advances that have allowed researchers to determine relatedness based on molecular markers, starting with the comparatively crude blood plasma precipitation assays in the mid-20th century to Charles Sibley's ground-breaking DNA-DNA hybridisation studies in the 1970s leading to DNA sequencing techniques. The results of these techniques caused revolutionary changes in the higher taxonomic categories (such as phyla and classes), resulting in the reordering of many branches of the phylogenetic tree.

For taxonomic categories below genera, the results have been mixed so far; the pace of evolutionary change on the molecular level is rather slow, yielding clear differences only after considerable periods of reproductive separation. DNA-DNA hybridization results have led to misleading conclusions, the Pomarine Skua - Great Skua phenomenon being a famous example. Turtles have been determined to evolve with just one-eighth of the speed of other reptiles on the molecular level, and the rate of molecular evolution in albatrosses is half of what is found in the rather closely related storm-petrels. The hybridization technique is now obsolete and is replaced by more reliable computational approaches for sequence comparison. Molecular taxonomy is not directly based on the evolutionary processes, but rather on the overall change brought upon by these processes.

The processes that lead to the generation and maintenance of variation such as mutation, crossover and selection are not uniform. DNA is only extremely rarely a direct target of natural selection rather than changes in the DNA sequence enduring over generations being a result of the latter; for example, silent transition-transversion combinations would alter the melting point of the DNA sequence, but not the sequence of the encoded proteins and thus are a possible example where, for example in microorganisms, a mutation confers a change in fitness all by itself.

Tree Species

Fagus Grandifolia

Leaves: elliptical to oblong, leathery, serrated, high content of tannin.

Branching: alternate; buds are long, like a lance, and have a golden colour.

Bark: smooth, light blue-gray (most distinctive feature).

Height: 70 to 80 ft.

Trunk Diameter: 2 to 3 ft.

Longevity: 300 to 400 yrs.

Tolerance: very tolerant

Range: eastern U.S., except Florida

Fun Facts:

- Wood used in inexpensive furniture and tool handles.
- Fruit (nut) is eaten by ruffed grouse, wild turkey, bobwhite, pheasant, raccoon, red and gray foxes, whitetail deer, cottontail rabbit, opossum, porcupine, gray, red and flying squirrels.
- Common to see initials carved in the smooth bark but this is not good for the tree.

American Chestnut

Castanea Dentata

Habitat: Northern hardwood forests, meadows, edges, pastures. Often grown as an ornamental tree because of its long feather-like flowers which appear in late spring and early summer.

Characteristics: At one time these trees attained a height of 100 feet and the trunks were 2-4 feet in diameter; however, a ubiquitous fungus (*Endothia parasitica*) precludes the normal growth of the tree. At present, most trees never grow taller than 25 feet.

Although it was formerly a dominant tree of the oak-chestnut forest, now only gray skeletons remain of this once majestic tree. A fungus disease which originated in eastern Asia was accidentally introduced into this country in 1904. By 1930, the disease had spread across the nation, killing just about all of our American chestnut trees. The few remaining trees growing today are root sprouts originating from the root systems of long-dead mature trees. Eventually the small trees contract the blight and die, too. However, as the older root sprouts die they are replaced by new shoots.

The wood was light and not very strong but very durable. It had a variety of uses ranging from furniture to railway ties. The chestnut was also one of our most beautiful ornamental trees and produced large crops of flavourful nuts enjoyed by man and wildlife alike.

American Hornbeam

Carpinus Caroliniana

Leaves are alternate, simple, double-toothed with equal leaf bases.

The American hornbeam is also known as the blue beach. It is a small, slow-growing, bushy tree with a spreading top of crooked or drooping branches. The trunk is more or less angular and is usually crooked. It is found mainly on moist slopes, along streams or on deep, moist woodland soils.

The leaf is dull-green above, pale green below and has hairy tufts at the base of the veins. It has depressed veins on the upper surface. The fruit occurs in spikelike clusters in a three-lobed bract, with a small nutlet attached to the outside. The leaf like bract may act as a wing in aiding seed distribution by the wind.

Fluted or irregular ridges like muscles on the smooth bluish gray bark extend up and down on the trunk of the tree and quite definitely identify the tree as the American hornbeam.

Basswood

Tilia Americana

Leaves are alternate, simple, double-toothed with unequal leaf bases.

The basswood is a large, wide-spreading, round-topped tree with dense foliage. The basswood has been widely planted for street trees. The tree blossoms in late May or June, and the blossoms are a favorite with bees, yielding large quantities of honey.

The leaves are rounded or heart-shaped, with toothed margins and with one side of the base less rounded than the other. They are dull green above and lighter green beneath. They vary quite widely in size from 4 to 7 inched long.

The twigs are smooth and light brown to gray in colour. The bark is light or silvery gray, and smooth or finely ridged on branches and young trunks. On older trunks the bark breaks into long, medium-narrow ridges and furrows, dark gray to almost black on the surface.

The tree has a very distinct fruit. It is an open cluster of hard nutlets borne on a stem which comes from the centre of a narrow, elliptical, leaflike wing.

Leaves: heart-shaped, fine-toothed edges, with uneven bases. Green and red give the leaves a purplish hue in fall.

Branching: alternate

Bark: grayish with tight even furrows

Height: 70 to 80 ft.

Trunk Diameter: 2 to 3 ft.

Longevity: matures in 90 to 140 yrs.

Tolerance: moderately tolerant

Range: eastern U.S., to southern Tennessee and western Virginia.

Fun Facts:

- Timber widely used for wood carving.
- Iroquois Indians made rope from the bark by soaking it in water for several days; it was then twisted into the desired form.
- Often has stump sprouts at base of tree.
- Knock on the trunk to hear a hollow sound.
- Edible buds.

Big Toothed Aspen

Populus Grandidentata

Leaves are alternate, simple, single toothed, triangle shaped. This is a fast-growing, relatively short-lived tree, sometimes reaching a to 80 feet and a diameter of 24 inches or more. It has a tall, straight bole. It is also known as Large-toothed aspen. The single leaves are 3 to 4 inches long, with a dark green upper surface and with large, coarse, irregular teeth. The buds are smaller than the cottonwood, more or less pale and downy rather than resinous.

The fruit is a catkin like the cottonwood fruit but the capsules are smaller and darker in colour, containing brown, fine, tufted seeds.

On branches and small trees the bark is light gray in colour. On older trees it is more yellowish or brown, breaking into dark gray to brownish-black shallow ridges near the base.

Leaves:

- usually larger than other Populus species
- coarsely toothed with a flattened slender leafstalk (petiole)
- young leaves whitish and woolly underneath

Branching: alternate

Bark: mostly smooth, olive green on young trees; later brown and furrowed

Height: 60 to 70 ft.

Trunk Diameter: 1 to 2 ft.

Longevity: 50 yrs.

Tolerance: intolerant

Range: north central and northeastern U.S. and southern Canada

Fun Facts:

- Produces more wood per acre than most other trees
- Used to make paper, pressboard
- Deer browse
- Sprouts from the roots (suckers)

Black Birch

Betula Lenta

This birch, also known as sweet or cherry birch grows in and along the slopes of deep, cool, moist ravines. It grows from

Maine, Quebec, and Ontario; to Delaware, Maryland, and Ohio; and in Georgia and Alabama.

The bark somewhat resembles that of wild cherry. Although the leaves of this and yellow birch are very similar, the leaves of the black birch are more heart-shaped at the base.

Oil of wintergreen is obtained by distilling the twigs and inner bark of this tree. One can readily detect the wintergreen oil just by smelling a freshly broken twig. Both birch beer and birch tea are make from black birch.

Black Cherry

Prunus Serotina

The black cherry is a valuable timber tree of medium size, reaching a feet and a diameter of 2 feet. It develops a large, rounded, oblong crown when grown in the open. It is found usually on rich, moist soils in mixture with other species.

The thick leaves are narrow and pointed at both ends. Leaves are finely toothed with teeth curving towards the tip; dark green and very shiny above.

The twigs are dark red-brown, sometimes grayish, with the distinctive taste of the cherries. The fruit occurs in purple to black clusters. On young stems the bark is grayish brown to red-brown, with prominent white spots. On older trunks it is dark reddish brown to almost black, and very rough, breaking into upturned, stiff thin scales.

Leaves: long, narrow, finely serrated; turn bright yellow (a few turn red) in the fall.

Branching: alternate

Bark: looks like burnt potato chips glued to the tree trunk

Height: 60 to 80 ft. (100 ft.)

Trunk Dumeter: 2 to 3 ft. (5 ft.)

Longevity: 150 to 200 yrs.

Tolerance: very tolerant

Range: eastern U.S., Arizona south through Mexico to Guatemala.

Fun Facts:

- Lumber is reddish in colour and used in furniture.
- Berries are eaten by songbirds, ruffed and sharp-tailed grouse, bobwhite, pheasant, raccoon, red fox, whitetail deer, cottontail rabbit, and gray squirrel.
- Scratch a young twig. How does it smell? (HORRIBLE!)
- After a branch is cut, a poisonous gas forms in the leaves just as they turn yellow.

Black Locust

Robinia Pseudoacacia

Leaves are alternate, compound, with thorns or spines.

The black locust's natural range extends from Pennsylvania southwestward to Alabama and westward to southern Illinois. The tree often suffers extensive damage from the locust borer insect.

The leaves are pinnately compound with 11 to 15 leaflets 1 to 1-1/2 inches long, rounded at both ends and with smooth margins.

The twigs are crooked and angular with short, stout, single, unbranched thorns, one-half inch long.

The fruit is a dark, red-brown, flexible pod 3 to 4 inches long, containing small, reddish brown beanlike seeds.

On young branches the bark is smooth and greenish to brown in colour. On older branches and trunks it is broken into a network of coarse, deep ridges and is gray to gray-brown in colour.

Leaves: pinnately compound; margin entire

Branching: alternate

Bark: mature trees reddish brown to nearly black, deeply furrowed. Bark is poisonous; cattle die from browsing on bark and children become ill from chewing on twigs or bark.

Height: 70 to 80 ft.

Trunk Diameter: 2 to 3 ft.

Longevity: 60 to 80 yrs.

Tolerance: intolerant

Range: east of Rocky Mountains

Fun Facts:

- Wood is resistant to rot and used to make railroad ties and fence posts.
- Fruit, pod (legume) eaten by wildlife.

Black Maple

Acer Nigrum

Leaves are opposite simple.

The black maple is very similar to the sugar maple and is often mistaken for it.

The leaves are three to five lobes, but with the lobes sharp pointed and not so deeply cut as the sugar maple. The leaves are dull green above and yellow-green below with soft hairs, especially along the yellow veins. The margins of the leaves droop, making them appear somewhat wilted.

The twigs are stout, light to dark gray, smooth and opposite on the stems. The fruit is a pair of winged seeds, ripening in autumn. The bark is very similar to the sugar maple, except that it is usually darker.

Black Oak

Quercus Velutina

Leaves are alternate, simple, lobed. Lobes have pointed tips. Fruit is an acorn.

The black oak is a medium sized tree, 60 to 80 feet tall and one to two feet in diameter, with a wide and irregularly shaped crown. It is found widely over the eastern part of the United States as far west as eastern Nebraska, Kansas, Oklahoma and Texas. It is found quite generally throughout Iowa on dry ridges.

The leaves are single, 5 to 7 lobed, bristle-tipped, 5 to 10 inches long and 3 to 6 inches wide. The lobes are sometimes shallow and sometimes deeply lobed, the shape varying greatly. They are dark green and shiny above and pale beneath, with rusty brown hairs in the forks of the leaf veins. The buds are distinctly angular. The twigs are stout and dull reddish brown to dark brown in colour.

The black oak's small, rounded acorn is often hairy or velvety. The acorn cap reaches one-half way up the acorn with a rough margin.

On branches and young trees the bark is smooth and dark brown. It is thick and black on older trees, with deeply furrowed and broken ridges. The inner bark is bright yellow and bitter to the taste, because of the tannic acid it contains

Black Walnut

Juglans Nigra

Habitat: Fertile soils in mixed hardwood forests. Also grows well in pastures, meadows, and slopes.

Characteristics: A major characteristic of this tree is its deeply-furrowed, black bark. Trees range in height from 70-150 feet and have a diameter of 2-4 feet. The compound leaves are between 1 and 2 feet long. This tree is prized for its beautiful wood and the tasty nuts which are avidly harvested in the autumn.

Walnut is unquetionably the finest wood in the world. Our forefathers sought it out and used it lavishly in their homes, barns and fences. The warm brown wood finishes beautifully. It is easy to work, yet durable. It shrinks and swells less than any other wood, which makes it valuable to cabinet makers and gunsmiths alike.

Practically everyone recognizes a walnut tree when the nuts are on it. Their distinct shape, pattern and smell are hard to miss. The stain the hulls leave on your skin when you hull walnuts is hard to get off; pioneers used this to dye their cloth. There are other distinctive characteristics to identify the tree in other seasons and at ages when it is too young to bear fruit. A sure method is to cut through a twig at an angle and check the pith. This pith is chambered—somewhat like a honeycomb. Only black walnut and butternut—a close relative—have pith like this. Walnut pith is brown and butternut is buff coloured.

The rather large, horse-faced leaf scars on the twigs, the large naked buds (no scales cover the embryonic leaves), and the smell of the bark and twigs are other easy ways to identify it.

High quality logs are made into veneer. Slices of wood 1/28th of an inch thick are glued to cheaper woods to make it econimically possible for all of us to buy walnut furniture and paneling. Demand far exceeds supply. We use both veneered and solid in our finest furniture. Its strength, stability and beauty make it unexcelled for gunstocks, too. Hickories and pecans are first cousins (botanically) to walnut.

Walnut grows in a variety of soils. However, it grows best on the deep, well-drained soils of north Missouri and on alluvial soils in the south. Usually it occurs as scattered trees or in small groves. Strangely, walnut roots transmit a growth-inhibiting chemical which keeps many other trees and plants and even its own kind from growing near it. Blue grass, however, thrives in the light shade of walnut.

It should be enough to have a tree with superb wood. Walnut has a double value, though, over a million dollars is paid each year for the nuts. The richly flavoured nut meats are used by bakers, candy and ice cream makers. The uses go even further. The hard shells are used as ornaments, and pulverized, they ares used to drill oil wells, clean jet engines and to make activated carbon (a type of industrial charcoal used in a variety of ways). During World War II, gas mask filters were made from this activated carbon. Wildlife loves the walnut, too.

Leaves are alternate compound with more than 11 leaflets.

The black walnut is one of our best known and most valuable trees. It is a large, straight-stemmed timber tree with an open crown. When grown in the open it is a short-trunked, low-branching, wide-spreading tree. Widely planted for its nuts, lumber and for ornamental purposes, it is found over the eastern United States as far west as Nebraska, Kansas and Texas. Growing chiefly along bottomlands, in coves and on lower slopes, it prefers a deep, rich, moist but well-drained soil.

The tree has large, pinnately compound leaves, 12 to 24 inches long with 15 to 23 leaflets. The leaf stems are covered with fine hairs, but are smoother than butternut. The leaflets are 2-1/2 to 3 inches long, yellowish green in colour, tapering at the end and toothed along the margin.

The twigs are brownish, stout, blunt and with prominent leaf scars. The pith is cream coloured and chambered, dividing into thin plates or segments.

The fruit is a large, rounded, brownish black nut with a hard, thick, finely ridged shell enclosing a rich, oily kernel. The kernel is edible and highly nutritious. The nut is enclosed in a solid, non-splitting husk, and is borne on the tree singly or in pairs.

The thick bark is dark brown in colour and divided by deep fissures into rounded ridges. It has a chocolate brown undercolour when broken from the tree.

Black Willow

Salix Nigra

Black Willow is also called Swamp Willow.

Distribution: Throughout the eastern United States, from the Atlantic coast west to Minnesota, Iowa, Nebraska, Kansas, Oklahoma and Texas.

The Tree: Black willow trees reach a height of 140 ft. (43 m), with a diameter of 4 ft. (1.2 m).

General Wood Characteristics: The sapwood of black willow is light tan, while the heartwood is pale reddish brown to grayish brown. The wood has no characteristic odour or taste. It is soft, but does not splinter when dented; uniform in texture; and weak in bending and crushing.

Like all willows, Black Willows aid in stream bank stabilization. Willows are pioneer, sun loving trees that are the first to become established on sand bars. Only black willow reaches commercial size. The bark of the root is intensely bitter and has been used as an ingredient of tonics to purge the blood. Usually black willow is a rather small tree, but under good conditions it may grow to 70 or 80 feet in height and two or more feet in diameter.

Willow trees are either male or female with flowers of different sexes borne on seperate trees. The female flower hangs in long clusters or aments. The seed is carried far and wide by wind and water.

Light and flexible willow wood is used for wickerwork furniture and basket. It does not split when nailes. Before the discovery of plastics, toys were made from willow. Artificial limbs, packing cases and some furniture parts are also willow products. Its warm brown tones and attractive grain make it a beautiful paneling wood.

Leaves are alternate, simple, long and thin.

The black willow is a large willow, attaining a height of 75 feet and a diameter of 2 feet or more. It has as irregular, open crown and frequently a cluster of main trunks.

The leaves are narrow (3/8 to 2/4 inch wide) and taper pointed, with long, sharp curved tips and finely toothed margins. They are shiny light green above, duller below and often velvety along the midrib. There are prominent leaflike stipules at the base of the leaf stem. The buds are reddish brown. The twigs are reddish or orange-brown, slender and brittle.

The fruit is a catkin of small, light reddish brown capsules, 1/8 inch long, with tufted seeds. On young branches the bark is yellowish to reddish brown. On larger stems and trunks it is almost black, broadly ridged and shaggy or scaly.

This is the largest and most picturesque of all our willows. It is only one of about 100 kinds of willows occurring in North America. All are very similar in appearance, and generally difficult to identify, even for experts.

Notice that the long, lance-shaped leaves are green on both sides. Often stipules occur at the base of the leaves. Stipules on vigorous shoots persist until autumn. Male and female flowers are borne separately in catkins on different trees.

The Patowatomi Indians made a fine scarlet dye from the roots of this tree. A tea brewed from the roots and bark was used as a substitute for quinine by the pioneers. Even today, willow charcoal is frequently used for manufacturing gunpowder.

Willow wood is soft, light, and weak but highly resistant to splintering and splitting. It is used for crates, charcoal, and artificial limbs.

Butternut

Juglans Cinerea

Leaves are alternate compound with more than 11 leaflets.

The butternut, also called white walnut, is found mostly on bottomlands and lower slopes on moist, rich soils.

The pinnately compound leaves are similar to black walnut leaves, being 15 to 30 inches long with 11 to 17 sharp pointed, oblong, finely toothed leaflets 2 to 3 inched long. The leaf stems and leaflets are velvety.

The large, lemon-shaped fruit has a shell which is very rough and sharply ridged, and the kernel is edible. The nut is enclosed in a yellow-green husk with short, rusty, clammy, sticky hairs. On young branches, the bark is smooth and light gray. On older branches and trunk it breaks into shallow, flat gray ridges which form a diamond-shaped pattern. The underbark is chocolate brown.

Chestnut Oak

Quercus Prinus

Habitat: Northern hardwood forests; prefers dry soils.

Characteristics: Trees attain a height of 60-70 feet, though taller trees are sometimes encountered. The trunks of mature trees are 3-4 feet in diameter; the leaves have either rounded or sharp teeth and are 4 to 9 inches long. The leaves are dull orange in autumn.

The strong, durable wood of the better-formed trees has the same uses as that of other oaks. The bark is rich in tannic acid, and was once collected for use in tanning leather.

This is occasionally a very long-lived tree. The famous Washington Oak, an exceptionally large chestnut oak which grew on the east bank of the Hudson River in New York, is estimated to have been 800 to 1,000 years old.

Crabapple

Malus sp.

Crab apples grow in woods, on the edges of fields, and on the edges of prairies. Their flowers are perfect for many insects. The tree yeilds fleshy, yellow-green pomes which are fragrant, waxy, and 1-1 1/2" in diameter. These are eaten by bobwhite, pheasant, and songbirds. The tree is of high value as nesting cover. The flowers are large and pink and emerge before the leaves unfold. The tree often forms thickets.

Description: Small thorny deciduous tree. Height 16m

Habitat: Common in Oak woods and hedges.

The Tree Year:

Flowers—May

Leaves—April

Fruit—Ripens in the fall.

Timber: Hard close grained wood.

Uses of Wood: Wood carving, inlay work, mallets, screws. Good firewood with pleasant aroma.

Food and Drink: The attractive small fruit are extremely sour but make good jelly by themselves or with blackberries or rowan berries. Crab Apple wine is reported to be potent.

Cucumbertree

Magnolia Acuminata

Habitat: Cucumbertree is a member of the deciduous community and is found at the higher elevations (3,000-4,000 feet) in the Appalachians. It prefers moist soils.

Characteristics: Attains a height of 40-70 feet; the trunk having a diameter of 1-3 feet. The leaves are 5-10 inches long and are similar in appearance to the other magnolias. The cucumber-like fruit is green when young and turns red when older; the seeds are bright red.

It is the most hardy of the seven tree-sized magnolias native to the United States. Cucumbertree grows singly or in small groups. The wood is light and weak, but it is durable and works easily. It is used for cabinet work, interior finish, boxes, and crates.

The fleshy, immature fruit looks like a small cucumber. As it matures it turns from green to a purplish-red colour. The leaves are quite large -from seven to ten inches long and from four to six inches wide.

Eastern Hemlock

Tsuga Canadensis

- The inside bark is purplish.
- The yellow-bellied sapsucker makes rows of small holes in trunk.
- It may live up to 250 years.
- It is the Pennsylvania state tree.

Other Common Names Hemlock, Canadian Hemlock

Needles are narrow, flat and soft, less than 1 inch long. Dark green above, with two white lines below. Like balsam fir, the needles on eastern hemlock seem to grow out from each of the twig. Each needle sits on a tiny stalk.

Twigs are slender, rough when needles fall. (Note: spruce twigs are much rougher than hemlock)

Buds are about 1/16 of an inch long, blunt, chestnut-brown in colour

Bark is reddish-brown on young trees, becoming dark and cracked on older trees. Quite often small dead branches stay on the lower trunk of the tree.

Cones In autumn cones are small, pale green at first, turning red-brown in late autumn, 1/2 to 3/4 inch long with thin, rounded scales.

Uses Hemlock is used for timbers and general construction, boxes and crates, railway ties and pulp.

The hemlock's small cones are among the smallest of the pine family. The with two white lines. These lines are actually four rows of white stomata (breathing pores) which are too small to be seen individually without a lens.

The hemlock is a tree that grows best in cool, moist ravines. It is distributed in such habitats throughout the eastern half of our state.

While once it was used only for pulpwood or as a source of tannin for tanning leather, now considerable lumber is manufactured from hemlock for cheap construction purposes.

The Indians used the moist inner bark to make a poultice for wounds and sores. Even today hemlock oil, distilled from the needles and twigs, is used in liniments.

Flowering Dogwood

Cornus Florida

Leaves are 3 to 6 inches long. The veins run parallel to the leaf margin. The mid-vein contains fine strands when pulled apart.

Leaves: oval to egg shaped with five or six pairs of veins

Branching: opposite

Bark: dark (red-brown), deeply checkered in an alligator hide-like pattern

Height: small bushy tree (taller in southern states: 30 to 40 ft.)

Trunk Diameter: 12 to 18 in.

Longevity: 60 to 80 yrs.

Tolerance: tolerant

Range: southern Maine to central Florida, west to eastern Texas

Fun Facts:

- Planted as an ornamental
- Wood used in industrial weaving looms and wagon hubs.
- Fruit and twigs of the tree are food for song and game birds, skunks, deer, rabbits, and squirrels.
- Large white pink-tipped flowers in the spring.
- Identification tip - gently tear leaf. Strings will appear between the torn halves.
- You can always tell a Dogwood by its bark!

Hawthorn

Crataegus sp.

Leaves are alternate, simple, with thorns or spines.

There are several hundred species of haws growing in the United States. It is extremely difficult to distinguish between the species.

This short tree, with thin, erect branching and a narrow, open crown, prefers the banks of streams and open hillsides.

The small leaves are narrowed or tapered at the base. The tips are round pointed and coarsely toothed, and the base is nearly smooth. The slender twigs are at first orange-green and smooth, later becoming bright chestnut brown and shiny. The older branches are ashy or reddish gray, with slightly curved spines of chestnut brown colour, 3/4 to 1-1/2 inches long.

The fruit occurs in few-fruited, small drooping clusters. It is dark red or rusty orange in colour with occasional dark dots and two or three seeds. On the trunk the bark is gray to dark gray-brown and breaks into narrow, flat ridges.

There are about 800 species of hawthorn in North America, of which more than 150 grow to tree size. Hawthorns are generally small, shrubby, thorn-bearing trees. Great individual variations complicated by frequent hybridization make accurate identification extremely difficult except for the professional botanist. Even botanists have not been able to agree on the total number of species found in North America. For these reasons no attempt is made here to differentiate between the many species of Crataegus.

Hawthorns are often found in pasture land, where their sharp thorns protect them from grazing cattle. Because of this they frequently are able to take possession of abandoned lands until shaded out by forest trees in succession.

Although the wood is of no commercial importance, many hawthorns are valued for ornamental plantings. The fruit, referred to as "haws," remain on the tree all winter, providing food for wildlife.

Mockernut Hickory

Carya Tomentosa

Leaves are alternate compound with five to nine leaflets.

The mockernut hickory is a tall, short-limbed tree with a narrow crown and upright branches. It is found quite generally in the eastern part of the United States as far west as Kansas and Texas.

The pinnately compound leaves have five to seven leaflets which are broader toward the tips and more wedge-shaped toward the base than the leaves of the other hickories. The leaflets are yellow-green and glossy above, pale to orange-brown beneath, with soft, velvety surface. They turn a beautiful yellow in the fall.

The oval, slightly pear-shaped fruit has a thick, strong-scented husk, which splits nearly to the base when ripe. The nut is reddish to dark brown, with a thick shell and small, sweet kernel.

On older branches the bark is light to dark gray and on the trunks it breaks into broad, shallow and somewhat platelike ridges that are not shaggy.

Pitch Pine

Pinus Rigida

Particularly common in poor, dry, sandy soils. Needles in bunches of three, and twisted. Twigs are unwhitened. Has the unusual ability to resprout at various locations along the trunk or at the base of the trunk, following fires. Cones are 4-8 cm long. Some individuals have serotinous cones.

Redcedar

Juniperus Virginiana

The redcedar is a small, hardy tree. It grows 40 to 50 feet tall. It is food for wildlife habitat and windbreaks.

This tree has two types of leaves - one dark green, very small and scalelike, clasping the stem in four ranks so that the stem appears square. The other kind, often appearing on young growth or vigorous shoots, is awl-shaped, quite sharp pointed, spreading and whitened beneath. The twigs are four-sided.

The berrylike fruits are about 1/4 inch in diameter and bluish to bluish white in colour. The fleshy part is sweetish and resinous, enclosing one or two hard seeds.

The bark is very distinct, being thin, reddish brown in colour. It peels off in long, shreddy strips.

Redcedar is one of our most durable woods. It is light in weight with an aromatic odour and a distinctly reddish to purplish red heartwood and cream-coloured sapwood.

Red Maple

Acer Rubrum

Leaves are opposite simple.

The red maple grows well on a wide range of soil types. It is fast growing and grows to a to 70 feet tall.

The leaves are arranged opposite on the twigs. The are broadly ovate with 3 shallow short-pointed lobes turning red, orange, and yellow in the autumn. The lobes sinuses are sharp, with toothed margins.

The fruit is a v-shaped samara (winged seed). The twigs are shiny red with numerous small lenticels.

Leaves:

- Three (sometimes five) main lobes
- Leaf margins are slightly serrated
- Notches between lobes are shallow
- Palmately veined
- Base of lobes are pointed
- Fall colours range from gold to deep purple and scarlet

Branching: opposite.

Bark:

- Twigs and buds are reddish in colour.
- Younger trees are smooth and light gray.
- Older trees have long, narrow, scaly plates with shallow fissures.

Height: medium sized tree 50 to 70 ft.

Trunk Diameter: 12 to 24 in.

Longevity: maturity reached at 70 to 80 yrs. (may attain an age of 150 yrs.)

Tolerance: intermediate

Range: eastern U.S.

Fun Facts:

- Wood sometimes used for furniture.
- Named for the bright red flowers that appear in the spring.

Red Oak

Quercus Rubra

The red oak is one of the largest and most important timber trees. One of the fastest growing of the oaks, it attains a to 80 feet and a diameter of two to three feet. It has a wide, spreading head with few far reaching branches. Found growing over southeastern Canada and the northeastern United States, it reaches west to central Minnesota, eastern Nebraska and Kansas. It is found over most of Iowa on a variety of soils, except on the drier clay uplands. It prefers moist, rich soils on north, east or northeast exposures.

The tree has a single, lobed leaf with seven to eleven pointed or bristly-tipped lobes. The lobe sinuses reach one-half way to mid-vein. The leaves are thin, firm, dull green above, yellow-green below, varying considerably.

The fruit is a large, broad, rounded acorn with a very shallow disk-like or saucer-shaped cup or cap.

The twigs are small, slender, greenish brown to dark brown. On young branches the bark is smooth and gray to greenish. On the trunk it breaks into long, narrow, shallow ridges flat and smooth on top. The underbark is light red.

Leaves:

- Have seven to 11 toothed lobes that are separated by sinuses extending about halfway to the midrib.

- Contain tannin, a substance that makes the leaves leathery and hinders decomposition.
- Dark red, fading to brown but may remain on the tree well into the fall.

Branching: alternate

Bark: reddish brown when young; mature tree is dark, furrowed and often laced with broad shiny strips (ski trails).

Height: 70 to 90 ft.

Trunk Diameter: 2 to 4 ft.

Longevity: 300+ yrs.

Tolerance: intermediate

Range: eastern U.S. except for the south Atlantic and Gulf Coastal Plains

Fun Facts: Acorns provide a food source for numerous birds and animals: Ruffed grouse, nuthatch, blue jay, wild turkey, red, gray and fox squirrels, bears, deer, raccoons.

Question: How many root beer barrels could you make from a red oak tree?

Answer: None - red oak has big pores, so all the root beer would leak out.

Red Pine

Pinus Resinosa

The red pine is a native of the Lake states and eastward throughout New England and southeastern Canada. It had not been planted widely in Iowa until the 1930's. Since then it has been planted quite widely for both erosion control and water conservation, and some for farmstead windbreaks. When growing under natural conditions, the red pine reaches a height of 90 to 100 feet and a diameter of 30 to 40 inches, with a tall, straight, clean trunk and an open, rounded picturesque crown. The tree gets its name from the bright orange-coloured or reddish bark, which divides into large plates as the tree matures.

Red pine needles are 4 to 6 inches long and in bundles of two. The dark green needles are soft and flexible. When bent sharply they snap or break cleanly rather than just folding over as do the needles of other pines. The cone is egg-shaped; 2 to 2-1/4 inches long. The cone scales are smooth and without spines. The seeds are eaten by songbirds and small animals.

Branching: Each year a pine tree produces a new whorl (circle) of branches.

Bark: reddish cast, breaking up into scaly plates

Height: 50 to 80 ft.

Trunk Diameter: 1 to 3 ft.

Longevity: maximum age is about 350 yrs.

Tolerance: intermediate

Range: southern Canada, lake states, and the northeast

Fun Facts:

- Most of the wooden telephone poles in Michigan and surrounding states are red pine.
- Used to make log cabins.
- During the Depression in the 1930s, millions of red pine plantations were planted by the C.C.C. (Civilian Conservation Corps.); that is the reason we see so many in Camp Conestoga.

Red Spruce

Picea Rubens

Habitat: Well-drained soils in mountain forests, usually above 3,500 feet. Also can be found along the borders of streams, swamps, and bogs.

Characteristics: Red spruce attains a height of 60-70 feet; the trunk of mature trees may have a diameter of 1-2 feet. The needles are dark green or yellow-green and are 1-2"-5/8" long. The cones are 1-1/4"-1-5/8"long.

Sassafras

Sassafras Albidum

Leaves:

- Margins are smooth
- Three patterns:

1. Three "fingers"
2. A "thumb and mitten" outline
3. Smooth, egg-shaped

Usually all are present on same tree

- Turn reddish-orange in the fall

Branching: alternate

Bark: mature tree reddish-brown and furrowed; when scratched, smells like root beer

Height: 10 to 50 ft.

Trunk Diameter: 2 to 12 in.

Longevity: 30 to 40 yrs.

Tolerance: intolerant

Range: eastern U.S.

Fun Facts:

- Tea may be made from boiling pieces of the outer bark from the roots.
- Sassafras oil is used in some soaps.
- Twigs browsed by rabbits and deer.
- Fruit eaten by songbirds, bobwhite, wild turkey, and bear.
- Wood was once used for barrels, buckets, posts, small boats, dugout canoes, and fuel.
- Spread from the roots.

Serviceberry

Amelanchier Arborea

Leaves are alternate, simple, single toothed, and oval or oblong shaped.

This tree is known by many names: serviceberry, Juneberry, shadbush, shadblow and May cherry. It is a small tree growing to a feet under favorable conditions. The tree flowers in the early spring, and has beautiful, delicate white flowers. It is desirable as an ornamental. The leaves are hairy when young and somewhat resemble cherry leaves in shape and outline. Leaves are dark red-brown as they unfold, but later dark dull green above, pale below and thick and firm. The slender twigs have long-pointed, yellowish buds. The tree has a smooth, dark ashy gray and matted bark.

Shagbark Hickory

Carya Ovata

Leaves are alternate, compound with five to nine leaflets.

The shagbark is the most common of the hickories and is an important timber tree with a narrow, open crown. It is found over the eastern United States and is quite general from bottomlands and moist slopes to the drier slopes and ridge tops.

The pinnately compound leaves are 8 to 16 inches long with five to seven dark yellow-green, broad oval leaflets with finely toothed margins.

The fruit is a brownish nut wit a thick shell and a sweet kernel, enclosed in a thick, splitting husk.

The twigs are smooth, or clothed with short hairs. The gray and very shaggy bark separates into long, narrow, hard, tough, loose scales, lightly attached to the tree.

This tree gets its name from the way the long, flat plates of bark break free on either end and curl away from the trunk, giving the typical shaggy appearance. Shagbark hickory has a compound leaf make up of five leaflets. The hard, bony-white nut, containing sweet, delicious nutmeat, is enclosed in a four-sectioned green husk.

No other hardwood has the combination of strength, elasticity, and toughness of hickory. The wood is used extensively for tool handles, athletic equipment, and for smoking meat. It makes the best firewood since it has the highest heat value of any wood.

Striped Maple

Acer Pensylvanicum

Habitat: Striped maple is very common in the Appalachians, especially at the higher elevations (3,000+ feet). It is found in woods, edges, slopes, clearings, and along roadsides.

Characteristics: **Striped maple has large, 3-lobed leaves; the bark is dark green with white stripes. Most striped maples are between 6-8 feet tall, though taller ones (20-30 feet) are often encountered.**

Sugar Maple

Acer Saccharum

Leaves are opposite simple.

The sugar maple (hard maple, rock maple) is one of our largest and finest forest trees, growing to a feet with a diameter of 2 or more feet. The tree produces a dense, round, compact crown when grown in the open and is used quite extensively as a shade or ornamental tree. In the fall the yellow, red and crimson colours of the leaves form a very showy and beautiful part of the landscape. It is the best of the maples for production of maple syrup and sugar.

It is found throughout southeastern Canada, the eastern United States and as far west as central Iowa.

The leaves are three to five lobed, but usually five lobed. The lobes are deeply cut with rounded divisions between the lobes, dark green above and pale green with a silvery cast below.

The twigs are opposite on the stem, smooth and gray to brown in colour.

Similar to other maples, the fruit is a pair of winged seeds about 1 inch long. The seeds ripen in the autumn.

On large branches and trunk the bark is light to dark gray, narrowly ridged with long, deep furrows, sometimes becoming scaly.

Leaves:

- Five main lobes.
- Margin (edge) of leaf is entire (not serrated).
- The margins of the central lobe are parallel.
- Notches between the lobes are moderately deep.
- Palmately veined.

- Base of lobes are rounded.
- Turn yellow, orange, and scarlet in the fall.

Branching: opposite

Bark: dark gray. As the tree matures, the bark becomes deeply fissured (looks shaggy).

Height: 60 to 80 ft.

Trunk Diameter: 1 to 3 ft.

Longevity: 200 to 300 yrs.

Tolerance: extremely tolerant

Range: eastern U.S. as far south as Tennessee

Fun Facts:

- Maple syrup is made from the sap.
- Lumber used for furniture and bowling alleys.
- It takes about 40 gallons of sap to make one gallon of maple syrup.

Tamarack

Larix Decidua

A small, slender tree also called the European Larch which rarely grows more than 45 feet tall. It has delicate, deciduous foliage.

The larch is not a true evergreen since it is one of the few conifers which lose their leaves in winter. It is a tall tree with a straight, central trunk and many spreading branches. It grows 40 to 80 feet tall.

The leaves are needlelike, but are flat, soft, slender, 3/4 to 1-1/4 inches long and borne in clusters on spurlike branches. In the spring the needles are bright green and very soft and flexible. In the fall they turn a dull yellow just before falling off.

The cone is small and nearly spherical, 1/2 to 3/4 inches long. The young cones are reddish to green in colour, turning brown as they mature often remaining on the tree for several years. As they mature and open in the fall, they liberate small, winged seeds.

The twigs are light brown to gray in colour, and covered with numerous tiny spurs or short branches. The rough bark separates on the surface into thin, reddish brown scales.

Tulip Tree

Liriodendron Tulipifera

Habitat: A principal tree of hardwood forests that grows in moist soil and is commonly found in pastures, meadows, farms, and along roadsides.

Characteristics: Tuliptree easily grows to a height of 100 feet and higher. The tulip-like flowers are a striking yellow-orange-green, and are a major nectar source for honey bees. The leaves are 6-10 inches long and have a distinctive appearance.

White Ash

Fraxinus Americana

Habitat: An upland tree found in the northern hardwood forests.

Characteristics: Trees reach a height of 70-80 feet and may have a diameter of 3-7 feet. The leaves of this ash and other ash species are opposite, feather compound. The leaflets occur in groups of 5-9.

Almost every American boy knows a great deal about white ash wood. He has felt the familiar "tock" and tingling jolt which accompanies the swinging of the ball bat made from white ash.

White ash grows over most of the state on moist, deep, well-drained soils in small isolated stands. Other relatives found in Missouri are blue, green and pumpkin ash.

Winged seeds are borne in clusters on the twig. These are spread by wind or water or a combination of both.

Leaves are opposite and compound with 5 to 7 leaflets which are oval and usually smooth-edged.

Smaller branches are stout, round and smooth. At the tip of the twig there are always three buds, the large, brown, pyramidal terminal and two small lateral buds. On older portions of the tree, bark is thick, dark brown, or grey and broken into flat ridges.

Because of its toughness, strength, and flexibility, the wood is used in the manufacture of athletic equipment, machinery, tools and furniture. It is a beautiful, light-coloured wood with a distincitive grain.

White ash seldom exceeds 100 feet in 1/2 feet in diameter. Because it is so hardy and has good form, it has been widely planted as an ornamental tree. It is rarely planted in forest plantations.

Ash is a lover of light, therefore it will not grow rapidly under shaded conditions.

White Oak

Quercus Alba

Leaves are alternate, simple, lobed; with rounded tips. Fruit is an acorn.

The white oak is one of our most important, largest, longest-lived and most valuable timber trees. It grows to 100 feet in to 4 feet in diameter. In the timber it forms a tall, straight tree, but in the open it is wide and spreading. Found over all of the eastern United States, it occurs on a wide variety of soils, but usually on upland clay soils.

The single leaves are 4 to 7 inches long and about half as broad, deeply divided into seven to nine rounded, fingerlike lobes. The young leaves are a soft, silvery gray or yellow to red when unfolding, later becoming bright green above and much paler be low.

The acorn is about 1 inch long, elliptical, and covered about 1/3 its length by a finely scaled, rounded cup.

The twigs are fine, and gray to green in colour. The bark is ashy gray to a very light gray and decidedly scaly. On older trunks it is somewhat ridged, but remains ashy gray and scaly.

Leaves:

- Have seven to nine lobes
- Smooth leaf margin (no tooth)

- Deep to shallow sinuses extending evenly to the midrib
- Contain tannin
- Turn orange-red, crimson, and red-purple, then fade to brown in the fall and may remain on the tree into winter

Branching: alternate

Bark: light ashy gray, mature trees - shallow and flaky

Height: 80 to 100 ft.

Trunk Diameter: 3 to 4 ft.

Longevity: 500 to 600 yrs.

Tolerance: intermediate, tendency to become more intolerant with age.

Range: eastern U.S.

Fun Facts:

- Acorns are food for wildlife.
- Wood is used for making furniture, flooring and pallets
- Has large pores, but plugged up by resin; wood could be used for root beer barrels.

White Pine

Pinus Strobus

The white pine is a large, straight-stemmed tree with a pyramidal crown. It grows 50 to 100 feet tall. It usually grows where there are limestone outcroppings and porous soils with good internal drainage.

The leaves are needles 3 to 5 inches long, bluish green on the upper surface, whitish beneath and occurring in bundles of five. The needles remain on the tree for two years.

The cone is slender, gradually tapering, 3 to 6 inches long with the ends of the cone scales prominently light tan to whitish and smooth. Cones are fully grown in the summer of the second season, opening to discharge seed that autumn.

On small branches and twigs, the bark is smooth. On older branches and the trunk, it breaks into broad, flat topped ridges and is dark gray.

Leaves: needles 3 to 5 in. long and in groups of fives

Cones: slender, thornless, 3 to 10 in. long

Bark: young stems thin, smooth and dark green; mature tree's bark is deeply furrowed.

Height: 80 to 100 ft.

Trunk Diameter: 2 to 3 1/2ft.

Longevity: maturity reached at 150 - 200 yrs.; some reported to be 450 yrs. old

Tolerance: intermediate

Range: southern Canada, Great Lake States, northeast and Appalachian Mountains.

Fun Facts:

- Michigan state tree.
- Tallest one in Marquette, Mich. is 180 ft.
- Largest of the northeastern conifers.
- Only conifer with five needles in a bundle, east of Rocky Mountains.
- One needle for each letter of its first name: W-H-I-T-E.
- Wood used in home construction.

Yellow Birch

Betula Alleghaniensis

Habitat: Commonly found at the higher elevations in the Appalachians (3,000-4,000 feet). Prefers moist woods and poorly-drained soils; ubiquitous in swamp forests and bogs. It is often seen growing near sweet birch, hemlock, and rhododendron.

Characteristics: A moderately tall tree that approaches 80 feet in height; the trunk having a diameter of 2-4 feet. The bark is yellow-bronze in colour and peels away in small curls. Broken twigs give off a fragrant wintergreen odour. In Ohio, the yellow birch is found in deep, moist, cool ravines in the northeastern quarter of the state and southward along the western border of the Allegheny Plateau. It generally grows in association with hemlock, sugar maple, and beech trees. The lustrous, smooth, silvery-yellow bark on the limbs and young trunks gives the tree its name. As the trunk grows larger, the bark breaks and rolls back in thin paper curls.

The twigs have a faint wintergreen scent like that of the black birch, with which this tree may be confused. However, the wintergreen scent of black birch twigs is very strong. About 75 percent of the lumber marketed under the name of birch comes from the yellow birch.

Tree Species

There are thousands of different tree species, hybrids, cultivars and varieties in the UK. However, according to some sources there are only 33 that are UK natives.

'Native' trees are defined as those that were able to migrate from mainland Europe without the aid of man. The period in which this refers to is after the ice age (9-10,000 years ago) and before the land connection with the continent was eroded, approximately 5,000 years ago. Those species listed below which are classified as native are coloured in green. There are a large number of trees that have been introduced since that period which have colonised very well in this country and are termed as 'exotic' species. Some of these are also included below.

Species Index

Native species appear in green:

Alder (*Alnus spp*) – General, Common, Grey

Alder Buckthorn (*Frangula alnus*)

Ash (*Fraxinus excelsior*)

Aspen (*Populus tremula*)

Beech (*Fagus sylvatica*)

Birch (*Betula spp*) – Silver, Downy

Blackthorn (*Prunus spinosa*)

Black Locust/False Acacia/Robinia (*Robinia pseudoacacia*)

Box (*Buxus sempervirens*)

Bracken (*Pteridium aquilinum*)

Cherries (*Prunus spp*) – Bird, Wild/Gean

Chestnuts (*Aesculus and Castanea spp*) – Horse Chestnut, Sweet Chestnut

Crab Apple (*Malus sylvestris*)

Dog Rose (*Rosa canina*)

Elder

Elm (*Ulmus spp*) – Wych

Eucalyptus

False Acacia/Robinia/Black Locust (*Robinia pseudoacacia*)

Firs (*Abies spp*)

Gorse (*Ulex europeaus*)

Guelder Rose (*Viburnum opulus*)

Hawthorn (*Crataegus spp*) – Common, Midland

Hazel (*Corylus avellana*)

Holly (*Ilex aquifolium*)

Hornbeam (*arpinus betulus*)

Juniper (*Juniperus communis*)

Laburnum (*Laburnum spp*)

Larch (*Larix spp*) – European, Japanese

Lime (*Tilia spp*) – Large Leaved, Small Leaved

Maple (*Acer spp*) – Field, Norway, Japanese, Silver

Oak (*Quercus spp*) – English, Sessile, Turkey, Holm

Pine (*Pinus spp*) – Scots, Corsican

Poplar (*Populus spp*) – Lombardy, White, Black, Aspen

Robinia/False Acacia/Black Locust (*Robinia pseudoacacia*)

Rowan (*Sorbus aucuparia*)

Sitka Spruce (*Picea sitchensis*)

Sorbus (*Sorbus spp*)

Strawberry Tree (*Arbutus unedo*)

Sycamore (*Acer pseudoplatanus*)

Tulip (*Liriodendron tulipfera*)

Western Hemlock (*Tsuga heterophylla*)

Whitebeam (*Sorbus aria*)

Wild Service Tree (*Sorbus torminalis*)

Willows (*Salix spp*) – Goat, Bay, Crack, White, Weeping

Yew (*Taxus baccata*)

Full Listings

Alders

The Common Alder (*Alnus glutinosa*) is a native deciduous tree of Britain which produces spectacular catkin displays in the early spring. It is a lowland species that loves to grow nears rivers and streams and in damp woodlands. The Grey Alder (*Alnus incana*) is planted in gardens and for shelter and can reach heights of up to 80ft. Although it prefers moist ground it does not tolerate water–logging in the same way as the Common Alder. It is a hardier species, often used on reclaimed industrial land. This species produces highly ornamental catkins before the Common Alder and before its first leaf flush.

According to ancient belief, evil lurked in the Common Alder. It was feared because its wood appears to bleed when cut. This caused superstition as it suggested the tree was the embodiment of a malign spirit.

The wood is yellow when seasoned, durable in winter and as it is easy to work, it is often used for clog making, shoe soles, charcoal for gun powder and broom handles. Dyes can be derived from the bark and fruit.

Alder Buckthorn: The Alder Buckthorn (*Frangula alnus*) is neither an Alder nor does it have thorns. However, it is often found growing alongside Alders in damp woods, in peaty soils and on raised bogs.

This shrub has many uses, and as the older name suggests ('Black Dogwood') it was often used to make butchers' spikes and skewers. Other uses for the shrub over the years range from explosives, laxatives, dyestuffs and purgatives to pea sticks.

Its autumn colours of yellow leaves and purple berries make for attractive, seasonal interest in natural woodland settings.

Ash: The Common Ash (*Fraxinus excelsior*) is a native tree to Britain and is found in the wild in moist woods, riverbanks and mixed broadleaved woodlands where it can reach heights of 135ft.

The pale, pliable and sturdy timber is used for a variety of things including tool handles, furniture, pegs, oars, hockey sticks, charcoal, tennis rackets and skis.

There was once a great deal of superstition attached to Ash trees; burning Ash wood was thought to remove evil spirits from dwellings, and sick children would be cured if passed through the cleft of an Ash tree. Scandinavians believed that the Ash was sacred and that Yggdrasil, the Tree of the World, was a giant Ash and that the tips of his crown resided in heaven and that his roots sought out the pits of hell.

The Weeping Ash (*F. excelsior 'Pendula'*) is a very attractive form, the branches of which sweep nearly down to the ground.

Aspen: The Aspen (*Populus tremula*) is a quick growing, deciduous tree and reaches heights of up to 30m. It is often found in woodlands, on wasteland and abandoned quarries and produces attractive catkins in spring. The Aspen quickly colonises new ground as it produces an abundance of suckers, although it is not a tree with a large population in Britain.

This tree is famous for the rustling sound of the leaves, hence its botanical name 'tremula', meaning that the tree can often be heard before it is seen. Country folk used to believe that the constant trembling indicated some secret grief or guilt for the Aspen. Some even believed it felt guilty for supplying the wood that was used for the cross of Christ.

As the tree does not grow very large, it is not an important British timber tree. However, before the extinction of the beaver in Scotland during medieval times, the Aspen was its staple food.

Beech: The Common Beech (*Fagus sylvatica*) is a native tree to southern England and is typical in size and form to forest trees. Its smooth, silvery trunk can be stunning and seen from miles around. It grows well on chalk downs or acidic soils and is an expert at seeking water. The Beech bears more shade than any other native tree except for Holly (*Ilex spp*) and Yew (*Taxus spp*) and although it can grow up under any tree, none other can grow in its shade. The tallest recorded tree in England is 150ft and Beech can reach 20 feet around their stems. The oldest surviving tree in Britain is said to be over 320 years old (although the average life span is around 250 years) and the tallest maintained hedge to be found is up to 105ft.

Beech play host to an unfortunately large number of wood decay fungi and other pests and diseases and mature to over mature trees can die quite quickly following infection from certain fungal species.

Beech wood is very pliable and bends easily, making it ideal for turning or furniture making.

Birch: There are two native British Birch, the Silver Birch (*Betula verriculosa syn B. pendula*) and the Downy Birch (*Betula pubescens*). The two species are fairly similar, having white bark and fairly open crowns, allowing plenty of other plants to grow under their canopy. This and the fact that they are comparatively small in height and spread makes them both ideal garden trees. However, the Silver Birch is thought to be more graceful, having pendulous branches. They are quick growing but short–lived, reaching heights of 60 – 80ft, although most are too short–lived to attain such growth. These deciduous trees are quick colonisers, with their seed usually the first to inhabit areas recently cleared of trees.

The sap of the Birch is rich in sugar and when tapped in spring it can be used to make wine by adding honey and can also be used as a shampoo (without the honey); while the oil derived from the bark can be used as an insect repellent. The wood of the Silver Birch is used for the backs of brushes and tool handles and the twigs in winter are used to make besom brooms for gardens and forest–fire beaters. The timber is not utilised largely as a crop as the wood is soft and rots quickly, although it is occasionally used to make furniture, tool handles and plywood. As firewood, it burns with a particularly bright flame.

Blackthorn: The Blackthorn or Sloe (*Prunus spinosa*) is a native, deciduous shrub or small tree of Britain that has virtually black, thorny branches. This formidable barrier of thorns can be used to protect plants or other property from humans and animals. In spring, it suddenly produces conspicuous white flowers prior to leaf burst and the birds nest in its protective shelter. During late summer, the birds quickly eat the berries, or sloe. Commonly used as hedging, wind breaks or barriers, this thorny shrub can grow to 16ft. It prefers chalky soil in a sunny position but is hardy and can cope with a number of environments. The sloe can be used to make preserves or sloe gin.

As the blossoms often appear in March during the period of cold east winds, a cold spring was traditionally known as a 'blackthorn winter'.

The Box (*Buxus sempervirens*) is most commonly seen as formal, clipped hedges or topiary. However, it is a native, evergreen species that once formed part of woodlands which are now very rare.

As a tree, it reaches heights of 35ft and has a small rounded form. The timber is hard, heavy, yellow and closely grained and is used for chess pieces, tool handles, drawing instruments and engraving blocks.

Bracken: The Bracken (*Pteridium aquilinum*) is an aggressive member of the fern family. It is a very successful plant living in a diverse range of habitats and once established, it can become the dominant species and be impossible to remove.

Cherries: There are two native species of Cherry to Britain, the Bird Cherry (*Prunus padus*) and the more common Wild Cherry or Gean (*Prunus avium*). Both species produce white flowers from March to April, while most of the common Cherries planted in parks, gardens and on streets are pink flowered exotic or cultivated species. The common, deciduous Cherries grow to around 60-70ft tall and can be found, more often than not, in woodland settings, parks or arboretums.

They are generally quick growing, and their timber can be used for fine furniture and veneers.

The fruits of these trees are very popular with birds in the summer, and although the fruits of the Wild Cherry are edible (used for syrups and cough mixtures), only birds are able to eat the Bird Cherry fruits, as the name suggests.

Chestnuts: There are two species of Chestnut common in Britain although neither are natives. The one that graces many parks, village greens or rectory lawns is the Horse Chestnut (*Aesculus hippocastanum*). This deciduous tree is of course famous with children and adults alike for the abundance of 'conkers' that are produced in the autumn. A cultivar is available (*Aesculus hippocastanum "Baumannii"*) that does not produce conkers but is similar to the Horse Chestnut in every other respect. The Horse Chestnut produces many foot–tall, 'candle–like' flowers of white, and the tree seeds itself quite freely in woodland settings. The oldest living specimen in Britain is 130ft tall and was planted in 1664.

The wood is pale cream or brown and although it is light it is weak and therefore not planted as a timber crop. However, it can be used for toys and for trays to store fruit as it is quite absorbent. It is light and easily shaped and until recently was used to form artificial limbs.

The other common Chestnut is the Sweet Chestnut (*Castanea sativa*) although it is from a different genus altogether. The most notable feature of this tree is the greatly spiralling bark, which extends up the trunk, although this doesn't occur until the tree is around 100 years old. Chestnut stuffing and chestnuts roasting on an open fire are a traditional winter delicacy in Britain, however, those sold in shops are rarely British. This is due to our comparatively cool summers, so they are instead imported from France and Italy.

Sweet Chestnuts are not great colonisers in this country as the seed does not germinate freely. The majority are planted specimens and are more commonly found in parks and large gardens. The timber is converted into panelling and beams as it resists the weather quite well, and is most often used for the palings of rough fencing.

Crab Apple: The Crab Apple (*Malus sylvestris*) is a native tree of Britain and the name is derived from the Norse word skrab, which describes its 'scrubby' nature. It can be found in mainly Oak woodlands and elsewhere throughout the countryside and can reach heights of around 35ft.

They provide valuable wildlife habitats for birds as the fruits ripen in the autumn and these are also used to make jams, wine and jellies for our own enjoyment.

The timber of the Crab Apple is reddish–brown, tough and finely grained. It is used for carving, mallet handles and fire wood.

It is from the Crab Apple that many of the eating and cooking apple trees we grow today have been cultivated from, although they do not carry the thorns of the wild native.

Dog Rose: The Dog Rose (*Rosa canina*) is a native rose often found in hedge banks and downs. It is a medium sized shrub with very prickly stems and white or pink scented flowers. It also produces bright red, egged-shaped fruits.

Elder: Common elder (*Sambucus nigra*) is found growing naturally in woodlands and hedgerows throughout the country. It forms a large deciduous shrub or a small tree and is often considered a weed because of its ability to rapidly colonise. However, the red autumn berries and the white spring flowers are highly valuable to wildlife and provide a splash of seasonal colour. Most parts of the elder are used in making medicines, drinks, food or as talismans, such as fly repellent concoctions of leaves, elderflower water against sunburn, elderflower or elderberry wine, elderflower fritters and elderwood and sap as charms against witches and witchcraft.

Elm: There are a number of species of Elm that grow in Britain, however the deciduous Wych Elm (*Ulmus glabra*) is the only undisputed native. It can be found in most parts of Britain and is known to seed itself abundantly every year in favourable conditions. The Wych Elm is supposedly non-suckering, a common trait of most Elms, unless suffering from stress. The tallest recorded tree in Britain was 100ft tall and 25ft round at the stem.

Unfortunately, Elm trees are prone to Dutch Elm Disease, a killer which has caused the decline and death of a very large number of Elm trees throughout Britain and Europe. However, the Wych Elm is the hardiest of species in Britain in respects to climate and the local environment and it tolerates pollution well. The name 'Wych' is Anglo-Saxon and means pliable, referring to the twigs and a horseman carrying a riding-switch from an Elm is thought to be lucky.

The timber of Elm is extremely tough and can be an attractive reddish-brown with some green streaks. As it is durable even when wet, it has in the past been used for underground water pipes. Nowadays, as it is nearly impossible to split, it is utilised for the keels of boats, for harbour works, seats of chairs, hubs of wooden wheels and also provides the traditional wood for coffins.

Eucalyptus: There are approximately 500 species of Eucalyptus (*Eucalyptus spp*), mostly evergreen, large and very quick growing trees which are typical to Australia. The bark of the trees generally peels in long strips, leaving behind differing colours of powdery surfaces (reds, whites, blues, yellows and oranges) and some are very aromatic. Some species of eucalyptus are classed amongst the largest trees in the world and their variety of amenity features exceeds them as ornamental trees.

While they are generally very tall trees (some species reach heights of up to 350ft, although they average between 170-250 feet), eucalyptus can be easily coppiced once they begin to outgrow their location, so that a multi-stemmed tree is very quickly produced. A single stem can then be selected to grow on and the others removed or it can be left to develop a tall, shrub-like form.

Eucalyptus are now being recognised world wide for their value as ornamental trees (attractive peeling bark with contrasting colours left behind, delicate evergreen leaves, a range of scents), their value as very quick growing timber and for the manageability.

In Australia, their timber is utilised for construction, pulping, flooring, bridge building, wharf construction, telegraph poles and railway sleepers.

Firs: The common firs (*Abies spp*) found within this country include the noble fir (*Abies procera*) and the grand fir (*Abies* *grandis*). Introduced to this country in the early 19th century, these evergreen trees are slow growing to start with although their growth rates turn moderate to very fast. They are generally found within mixed conifer plantations and sometimes as specimen trees in parks and gardens. The grand fir is quite shade tolerant and can therefore be planted underneath already established trees. Both trees prefer abundant quantities of rainfall, although the grand fir will tolerate drier conditions. Both trees are native to North America.

Gorse: Common gorse, or furze (*Ulex europeaus*) is a common shrub found on heaths and rough grassland throughout the British Isles. It is a spiky, evergreen shrub which has the ability to spread and colonise large areas of land. It produces profuse, yellow flowers in the spring which provide a colourful scene for the evergreen backdrop.

Guelder Rose: The Guelder Rose (*Viburnum opulus*) is a native deciduous shrub of Britain and it is highly regarded for its ornamental qualities. In the spring it develops beautiful diamond clusters of white flowers and in August the berries turn red at the same time as the leaves start to turn burgundy. It can attain heights of 13ft but is normally much smaller.

Formerly, it was known as the Water or Swamp Elder, which refer to its Elder–like berries and for its love of growing in damp areas. It is often found in the Fens and in boggy areas, alongside Alder Buckthorn, Sallows, Birch and Alder. The leaves, bark and berries of this attractive shrub are all poisonous.

Hawthorn: The Hawthorn, May or Quickthorn (*Crataegus monogyna*) is a native, deciduous shrub or small tree of Britain and is most notable as a hedgerow species. Hawthorns provide ideal wildlife habitats for birds and a number of insects, and act as stock-proof barriers due to their thorns and dense growth habit as hedges.

The Hawthorn is fast growing and sturdy. Grown as a tree it can often be found on woodland edges or rides and in the open countryside, reaching heights of up to 35ft. They produce attractive white flowers in the spring and red berries in the autumn.

During the 16th and 18th centuries, land enclosures prompted the planting of thousands of hedges to surround estates. Unfortunately, a lot of these are now being destroyed due to mechanised farming. The spikes, which cause formidable barriers can now be replaced with barbed wire or electric fences.

Early Christians associated the Hawthorn with Joseph of Arimathea. He was the owner of the tomb in which Jesus was placed after the Crucifixion and later he was supposed to have planted his hawthorn staff into the ground at Glastonbury. This thorn apparently flushed new growth and was labelled a 'Holy Thorn'.

Pagan and medieval rites used the Hawthorn to symbolise the coming of summer, probably because of the early blossoms it produces. In some areas, it is believed that the destruction of a Hawthorn brings about doom and to have blossoms inside a dwelling invites disaster.

The Midland Hawthorn (*Crataegus laevigata*) is also a native species and very similar to the Common Hawthorn, although the leaf shape, number of seeds and flower stems is different.

Hazel: The Common Hazel (*Corylus avellana*) is a deciduous native to all parts of the U.K. except the Shetland Isles. It grows most commonly on chalk, limestone, neutral and mildly acidic soils. Hazel is often coppiced and grown for its timber and nut harvest, but even so is only a small tree, rarely exceeding 15m.

Hazel rods have been used since prehistoric times to weave a number of products, including baskets, wattle and daub building frames, fencing and hurdles. To provide continual rods, the Hazel is often coppiced so that the stool, or stump, produces an abundance of new shoots instead of a single stem.

Hazel rods and living trees are often used to form hedges in the country, and are sometimes mixed and interwoven with laid Hawthorns hedges to produce a dense barrier.

Holly: The Holly (*Ilex aquifolium*) is an evergreen, native tree or shrub to Britain and is found in most environments, including harsh ones, although it does not do well in wet soils.

The red berries produced in autumn are a great source of food for birds and provide a view of colourful, seasonal interest.

It was at one time superstition that to cut down Holly would be unlucky and because of this, many Hollies still survive today and can be found growing within maintained hedgerows.

Sometimes, Hollies are clipped and shaped as ornamental trees and topiary and they are also ideal for the creation of dense, formidable hedges.

The wood is strong, heavy and white and has been used for carving and woodcuts.

Hornbeam: The Hornbeam (*Carpinus betulus*) is another British native tree that can be found in its wild state in Oak and Beech woodlands at a height of up to 80ft.

In the spring, the tree is adorned with yellow male catkins and in the autumn the leaves turn from green to yellow to gold to orange–brown. With age, this deciduous tree develops a deeply fluted and twisting stem, making this an attractive amenity tree.

It also makes an excellent hedge as when it is clipped it becomes quite dense and it holds onto some leaves during the winter months.

The wood is tough and is thought to be one of the hardest in Europe, so it is therefore also quite durable. It is used for making butchers' blocks, cattle yokes, mill wheels, piano keys, charcoal, mallets, skittles, bean sticks and as a fuel.

The Hornbeam was named after the wood's strength (Old English; horn: hard and beam: tree).

In its native state it is not very wide–spread in Britain, although it has been planted in parks, woodlands, arboreta and large gardens throughout.

Juniper: The Juniper (*Juniperus communis*) is a small, evergreen, native conifer tree or shrub of Britain and can reach heights of around 20ft. It produces dark, purple berries which ripen to provide the flavouring for gin or for culinary purposes.

The oil from the leaves was used in Ancient Egypt as part of a concoction for preserving mummies. Junipers were sacred to medieval Europeans as their burning branches and leaves kept evil at bay.

The Juniper grows in a variety of environments, from the chalky southern downs to the acidic pine forests of Scotland.

Laburnum: While laburnums (*Laburnum spp*) are not native to the UK, they are a common feature of many small and large private gardens. The characteristic golden bark, long trailing yellow spring flowers, dark heartwood and light sapwood make this a prized ornamental and wood turning timber tree. Unfortunately, all parts of this small tree are poisonous and planting should be avoided particularly in areas frequented by children.

Larch: None of the Larch species are native to Britain but one, the European Larch (*Larix decidua*), is particularly common and the one you are most likely to see. It is a deciduous conifer with a distinctive flush of bright apple green foliage in spring, which darkens during the summer to become golden brown in autumn. A very tall–growing tree, it can attain a height of 150ft, especially when grown in plantations that have been left to over mature or in sheltered, steep sided valleys.

Commonly planted in forests, gardens and shelter belts, this tree is notable for its stunning autumn colours and general seasonal interest.

The timber of the European Larch is of high quality due to its rust–coloured heartwood and is often used to make stair–cases, light furniture and wall panelling. As it is quick growing, it is also useful as a nurse crop in hardwood plantations as it can provide shelter to the more slow growing species.

While it is quick growing, it is however a short lived species and is susceptible to a number of ailments such as canker and it suffers under spring frosts.

The Japanese Larch (*Larix kaempferi*), as the name suggests, is a native of Japan but it is planted in considerable numbers as a plantation tree throughout hilly, western regions of the British Isles. It is also commonly planted as an 'edge' tree to new woodlands as its quick rate of growth makes it an excellent 'nurse' tree.

It is a tall, deciduous conifer that is relatively tolerant of polluted air and often planted on reclamation sites, particularly old coal tips.

Its orange twigs in the winter and its autumn colour also make this tree an ideal seasonal interest, ornamental tree. In Japan, it is a common bonsai species and they are believed to symbolise mortality and the changes that come about by the seasons.

This tree is quickly becoming more preferable than the European Larch, as it is not susceptible to cankers and grows at a faster rate.

The Japanese Larch is sometimes planted as a fire break amidst plantations of more inflammable conifers, as its wood smoulders rather than burns. However, as firewood, larch can be uncomfortably sparky. The wood can be utilised to make fencing and for structural work, although it has to be carefully sawn as the planks can easily twist.

Lime: There are two native, deciduous species of Lime (*Tilia spp*) in the British Isles and various hybrid forms between these two as well as some introduced exotics. The Large Leaved Lime (*Tilia platyphyllos*) is found naturally on the limestone soils of the Wye Valley and in South Yorkshire. However, it is also a popular street and parkland tree. The Small Leafed Lime (*Tilia cordata*) can also be found on base rich soils, particularly limestone cliffs. It grows naturally as far north as the Lake District but has been planted in parks, avenues, streets and gardens north as far as southern Scotland.

In high summer it is covered in scented yellow flowers which are very popular with honey bees. This tree is noted for supporting large numbers of aphid insect populations on its heart shaped leaves which are famous for their secretion of honeydew. Others that make a regular feast of the Lime tree are moths, caterpillars, cattle and deer and as a consequence, are rarely found in the wild.

Lime are long lived and hardy trees and often produce prolific sprouting, or 'epicormic growth' from the base. However the Large Leaved Lime does not produce this growth which makes it a more ideal street tree than the others.

Long surviving Small Leaved Limes can be found at Hampton Palace, that were planted 350 years ago by Charles I.

It was once believed that the Lime was a holy tree and it was widely planted in villages by peasants to deter evil. Due to its dense and wide spreading canopy it has always been a favourite shade tree, even as far back as the ancient Greeks and Romans.

The wood of the Lime is soft and even grained, making it popular with carvers and wood turners and as it does not warp easily, the keys of pianos and organs are also made from its timber.

Maples: The native Maple of Britain is the Field Maple (*Acer campestre*) and although it is most notably used in hedgerows it can grow to heights of 60ft or more and provides a wildlife–rich understorey in some woodlands. One of the oldest recorded trees dates from around 1700, although old specimens are not often found.

There are 150 recognised species of Maple (*Acer spp*) and small, colourful, Japanese varieties are commonly planted in gardens.

One of the larger specimens of Maple found is the Norway Maple (*Acer platanoides*) which grows to around 80ft, has slightly snake-barked trunks and is common as street or park tree. This tree tolerates pollution and grime well, making it an ideal city tree. The yellow flowers produced in the spring are popular with bees but unfortunately the trees are also hounded by grey squirrels, which strip off the bark for food as well as the underlying sweet sap. In the autumn the red, yellow and scarlet-brown leaves make it an attractive ornamental tree.

The Silver Maple (*Acer saccharinum*) can grow to around 70ft and prefers moist soil near to water and is a hardy, light-loving species. The delicately shaped leaves, their silvery undersides and the autumn colours make it an ideal ornamental, seasonally interesting tree.

Oak: There are two deciduous native Oak trees to the British Isles; the English Oak (*Quercus robur*) and the Sessile Oak (*Quercus petraea*). However, there are 37 species listed as being hardy to the climate of Britain. Oaks are a very important tree in Britain for their timber, their wildlife habitats, their extensive life spans and their grace and beauty.

The Sessile or Durmast Oak is dominant in northern and western areas. It prefers light soils, as it will be out competed in heavier clay and loam soil by the English Oak. The tallest Oak in Britain was a 142ft specimen in Herefordshire. Oak wood has for many years been used to produce charcoal as it burns steadily, so from early times it has been an important timber crop. Unfortunately, the large demand for this charcoal and the slow growth of this tree has meant the demise of large tracts of Oak woodlands. The oak bark has also been used to produce tannin, which is utilised in the tanning process for leather.

The English Oak predominates in the south of the country where soil conditions are more to its liking. It is a commonly planted tree in parks and woodlands where many giant veterans have survived to live almost 1000 years. Druids in Celtic Britain thought of the Oak as a sacred tree and the mistletoe that was more common then was gathered and used in secret rites. It is a very sturdy tree and is used in construction, panelling and furniture and the acorns are still sometimes used as animal fodder today.

The most commonly found exotic or non–native Oaks in Britain are the Turkey Oak (*Quercus cerris*) from Southern Europe and the Holm Oak (*Quercus ilex*). The Holm Oak is evergreen and often planted in parks and gardens and it does reproduce naturally in this country. The Turkey Oak however, has naturalised far more successfully and is spreading like a native throughout the countryside.

Pines: The Pines are a large and diverse group of evergreen conifers with over eighty species in the northern hemisphere and across the equator to Java. The Scots Pine (*Pinus sylvestris*) is the only native to Britain, however, there are a large number of other species commonly planted as garden, park and forest trees, many of which grow to a similarly large size. Pines are important wildlife trees, particularly in the north, where they provide some of the last outposts for the endangered red squirrel in the UK. The Scots Pine is distinctive because of its ornamental, bright orange bark in the upper crown and pale greenish–brown needles. It grows rapidly and can reach heights of around 115ft. It rarely exceeds a life span of 250 years except in exposed areas, where its growth is stunted and thereby slowed.

The native Scots Pine woodlands are now a rarity, as the timber they produce is of a high quality and in great demand. It is a tree commonly found in the Highlands and the lochs and rivers were ideal for floating the logs to be processed. The timber is used for buildings, ships' masts, charcoal, flaming torches, chipboard, boxes, fencing, telegraph poles and furniture and the tree has also proved a good source of turpentine, resin and tar.

The Corsican Pine (*Pinus nigra var. maritima*) is an evergreen conifer which is native to Corsica and Southern Italy. It is widely grown in parks, gardens, church yards and as shelter belts or commercial plantations in the UK. As a plantation, they provide very suitable habitats for deer and birds. It is found on most soil types including heavy clays, sand, chalk and shallow peats.

The Corsican Pine is a large tree reaching an average of 115ft in height, the tallest recorded in the UK being 145ft tall. However, the trees need plenty of light to grow quickly and in plantations they need to be thinned out carefully.

The timber is used to make pit props and in general building work. As plywood, thin strips of wood are shaved from the log in a spiral and glued together to form layers. The opposite angled grain in successive layers provides its strength.

Poplars: There are a number of species of Poplar that grow in Britain and the majority of these deciduous trees take up vast amounts of water to assist in their rapid growth rates.

Some of the most commonly recognised are the Lombardy Poplar (*Populus nigra 'Italica'*), which are very tall, dense, columnar trees often used as wind breaks and screens. They can reach heights of up to 120ft and tolerate soot and smoke very well.

The White Poplar (*Populus alba*) is notable for the white, felty undersides of its foliage and the whispering they cause in the breeze. In Greek legend this tree was originally black. Apparently, Hercules wore a garland from the White Poplar in a battle with Cerberus, the guardian of the underworld and his sweat bleached the garland white. This tree can reach heights of around 110ft, it suckers freely and tolerates pollution and salt, so it flourishes on road sides and near the sea.

The Black Poplar (*Populus nigra var. betulifolia*) is a native timber tree to Britain. It flourishes on fertile soils beside water and tolerates pollution well. It was often planted as a screen and its timber is utilised for baskets for fruit, clogs, matches and matchboxes. Unfortunately, the numbers of living trees surviving modern Britain is frighteningly low and proactive groups have formed around the country to help protect and regenerate this native tree by vegetative propagation.

Robinia: The Robinia (*Robinia pseudoacacia*) is also known as the False Acacia or the Black Locust. This tree can produce mean spikes along its trunk and within the branches as a form of protection against browsing animals. It is a fast growing, deciduous tree that reaches heights of up to 85ft. It can be an attractive ornamental tree with its delicate leaf formation, twisted trunk, white, sweet-scented hanging flowers and open crown. The variety *R. pseudoacacia 'Inermis'* does not produce the spikes and the variety *R. pseudoacacia 'Frisea'* has attractive honey-golden leaves.

Rowan: The Rowan or Mountain Ash (*Sorbus aucuparia*) is a deciduous native to all parts of the British Isles. It is generally a small tree rarely exceeding 50ft and often inhabiting harsh environments such as moorland and heath, although it is by no means restricted to these areas. It is distinctive in autumn due to its large clusters of red berries and is often planted in arboretums, streets and gardens.

From ancient times this tree has been associated with witchcraft and the name is thought to mean 'charm', from the Norse word *runa*. It apparently wards off witches and was often planted in churches and outside of dwellings. Sprays of Rowan leaves were traditionally hung over doors on May Day to thwart evil and wells dressed with Rowans also kept witches away. The alternative name of 'Mountain Ash' reflects the altitudes in which this tree grows, although its compact form makes it ideal for many planting sites. It is also very hardy and its branches can withstand frosts more than most trees.

The berries can be made into a jelly, which is a favourite accompaniment to game dishes and were also made into a drink to prevent scurvy, as they are rich in vitamin C. At one time, they were used as bait to trap thrushes, redwings and fieldfares by bird catchers.

The strong, flexible, yellowish-grey timber was once widely used to make the handles of tools, for small carvings and occasionally in place of Yew for making long-bows.

Sitka Spruce: The Sitka Spruce (*Picea sitchensis*) is one of the largest spruce trees and is used widely as a plantation tree. This tree is noted for its particularly prickly needles, which can make the tree painful to handle, although the advantage is that it is not revered by browsing animals.

This tree colonises easily and is widely planted. It can withstand severe wind exposure, including salty conditions and is very cold-hardy (except for late frosts). It requires plenty of light and will not tolerate shade. It is favoured as a plantation tree for its ability to establish on sites that range from fertile and sheltered to infertile and exposed.

The timber is prized, as the tree is quick growing and is suitable for building construction, pallets, paper pulp, particle board and plywood. However, this is one of the few spruce that is not used as a Christmas tree.

Sorbus: Sorbus is the common genus name for a very large number of species of tree across the globe. The most common and one of only 3 native to Britain is the Rowan (*Sorbus aucuparia*). However, a great many other species are planted as street trees or in gardens and parks for a number of different attributes. Most have very striking berries, which are a great attraction to wildlife, some form compact, columnar habits which make them ideal as street trees and others have stunning displays of autumn leaf colour for seasonal interest.

Strawberry Tree: The Strawberry Tree (*Arbutus unedo*) is a native evergreen tree of the UK that was naturally found in thickets, woods and dry rocky places throughout Britain, although it is generally only found in the wild today in Ireland. Nowadays, it is planted in arboretums and occasionally in parks.

This highly ornamental tree generally only reaches heights of 40ft and sports dark-reddish peeling bark, shiny dark-green evergreen leaves, scarlet fruits and simultaneous flowers.

Sycamore: The Sycamore, 'Great Maple' or 'Great Plane' (*Acer pseudoplatanus* is not a native to the British Isles but has naturalised to such an extent that it has spread abundantly throughout the country. It is a large deciduous tree growing to 115ft or more and sheds distinctive, winged seed-pods in the autumn.

In some woodlands it is considered a pest as it out–competes the native species, but the Sycamore has a lot to offer both to the commercial forester and the wildlife enthusiast. It is unique as a hardwood in its ability to withstand salty winds, and is often used as a shelter belt in these harsh conditions.

It is quick growing and by the age of 60 it can be felled for its timber, which is creamy white, does not warp and is easily worked. It if often used to make furniture, for musical instruments and for veneers, particularly those unique specimens that have a highly prized 'ripple' effect.

The Sycamore is sometimes known as the Martyrs' Tree as it was in Tolpuddle in Dorset, under one such tree, that the Tolpuddle Martyrs met in 1834 to form a society to fight starvation and poor wages. Unfortunately, they were caught and sentenced to seven years transportation to Australia, however, they were pardoned after two years and returned home.

Tulip: The tulip tree (*Liriodendron tulipifera*) is a moisture–loving ornamental tree that grows to heights of around 35 metres. In sheltered, southern climates, this tree can grow very rapidly, reaching almost half of its height in 20 years, although it has been known to live for over 280 years.

The tulip tree has a high amenity value, particularly for its honey-yellow, unusually-shaped leaves in autumn. The timber has always been highly prized, from the North American Indians who used the hollowed out trunk to make canoes, to modern times where it can be used to make kitchen furniture.

Western Hemlock: The Western Hemlock (*Tsuga heterophylla*) is a tree native to North America and found in particularly large numbers on the west coast. However, after its discovery by the British, a number have been planted throughout the UK in gardens, shelterbelts, estates and plantations.

This is a fairly quick growing tree that reaches heights of up to 260ft in ideal conditions of high humidity and high rainfall. This attractive, evergreen conifer can grow under deeper shade than any other conifer and its viable seed can be produced in copious quantities. The leading shoot of this species grows with a droop, so that it can grow through overhead branches without damage.

Whitebeam: The Whitebeam (*Sorbus aria*) is a native deciduous tree found from the lowlands to the mountains and it thrives on chalk and on limestone. In its natural state is can reach heights of around 80ft although its form is quite upswept, making it ideal as a modern street tree. This is supported further by its ability to withstand high levels of pollution and its high amenity value.

This tree has a variety of ornamental qualities, its silvery leaves opening in spring and persisting through the summer, red berries in the autumn and golden foliage prior to leaf fall.

The timber of the Whitebeam is very tough, heavy and hard and was once used for making the cogs of some early machinery. It can also be used to make handles or spoons and the fruit is often made into a jelly, traditionally eaten with venison.

Wild Service Tree: The Wild Service Tree (*Sorbus torminalis*) is another native tree to Britain although it is one of the rarest. However, there were times when this tree was more common but it is thought to have succumbed to the fires of charcoal burners since pre–Roman times.

The tree does not germinate freely and grows quite slowly, which is another reason for its lack of distribution, although it can occasionally be found in Oak, Hornbeam or Beech woodlands. This is an attractive tree; the bark peels off into strips leaving a chequered effect and the purple, autumn colour of the maple–like leaves can be stunning.

Willows: There are a large number of Willows that can commonly be found in Britain. Willows are generally thought to be quick growing and love an abundance of water, however there are many types of Willows with different needs and some prefer moist, well drained soil rather than wetlands and others grow slowly and stunted as alpines.

The Goat Willow, or Pussy Willow (*Salix caprea*) is a native, deciduous species and its golden, male catkins are often used in churches on Palm Sunday, but the well known catkins used in flower arrangements and indoor decoration are from other closely related species. Their flowers provide early food for bees, and as the tree is also pollinated by wind, it is an easy coloniser of surrounding ground, particularly damp places. The bark contains tannin and also the drug salicin, from which asprin is derived. The wood is very soft and was once made into clothes pegs and hatchet handles.

The Bay Willow (*Salix pentandra*) is another native, deciduous species and it can generally be found in its natural state on riverbanks and within meadows. It produces attractive male and female catkins in the early summer and its dark green glossy leaves have a reflective quality near the water.

The Crack Willow (*Salix fragilis*) is again a native, deciduous species which also prefers river banks. Its twigs are very brittle and it uses this as a form of vegetative propagation – the twigs snap off, travel down the river and once landing on ground again will take root and produce new trees. The wood can be utilised to make quite decent charcoal and is popular with artists for drawing. A purple dye can be derived from the roots and at one time this was used to decorate eggs at Easter. The Crack Willow is quite tolerant of pollution and salty air and therefore proves invaluable in coastal areas, particularly as the roots hold potentially eroding soils together.

The White Willow (*Salix alba*) is a quick growing, deciduous native tree which frequents damp areas, particularly near river banks and lowlands. This tree can reach heights of around 80ft, although traditionally it was pollarded at head height so that a regular supply of poles was available that was out of reach of grazing animals. The poles were used in fencing and other woven products as well as for firewood. Again, as a water edge species, this tree is invaluable as its roots hold together the soils that may otherwise be eroded by water and other elements. The timber of this tree has been used for flooring, the bottoms of carts and is most famous for the production of cricket bats in the south.

The most memorable Willows to most people are the Weeping Willows (*Salix chrysochoma*) which produce large, pendulous 'umbrellas' for crowns with branches often sweeping down to the ground.

Yew: The Yew tree (*Taxus baccata*) is an evergreen native to Britain that is slow growing and long lived and specimens have been found that are over 5,000 years old and up to 80ft tall. These are very commonly found in church yards and although some assume that the Yew was a favourite tree to plant on sacred ground, it is more likely that churches were built near to existing yews. It is thought that this tree was prized as sacred by the pagans and that when their religion was overtaken by the Christians, pagan areas of worship were eradicated and built upon by the newcomers.

However, there are few long–lived trees remaining in Britain today, and this is partly due to the timber being valuable in the Middle Ages for making long bows, as the wood is quite strong (heartwood) and elastic (sapwood). More recently it has been used for cabinetwork and wood sculpture and the dense growth habit forms ideal hedges.

Yews prefer to grow on chalky soils and can tolerate pollution well. They can grow under shade and can themselves form a dense shade so that very little will grow under their canopies.

Some elements of the Yew tree are poisonous, including the bark, needles, shoots and the seed.

2

Precautionary Approach to Species

Because of the high probability that impacts of species introduction be of irreversible and unpredictable impacts, many species introductions are not precautionary. Therefore, a strictly precautionary approach would not permit deliberate introductions and would take strong measures to prevent unintentional introductions. Recognizing the difficulties with introductions, the objectives of a precautionary approach to species introductions in relation to capture fisheries should be to reduce the risk of adverse impacts of introductions on capture fisheries, to establish corrective or mitigating procedures (as in a contingency plan) in advance of actual adverse effects, and to minimize unintended introductions to wild ecosystems and associated capture fisheries.

In relation to aquaculture, experience has shown that animals will usually escape the confines of a facility. As a consequence, the introduction of aquatic organisms for aquaculture should be considered as a purposeful introduction into the wild, even though the quarantine/hatchery facility may be a closed system.

Introductions and transfers (hereafter referred to as introductions) are an effective means to increase protein, generate income and provide employment. However, some intended and many unintended introductions may result in significant and serious impacts on capture fisheries. The numbers of unintended introductions, for example by means of ballast water, greatly outnumber those purposefully introduced for capture fisheries.

For a precautionary approach to fishery management, irreversible changes in the time scale of human generations and other undesirable impacts should be avoided, taking into account uncertainty. Species introductions, either purposeful or unintended, may have such undesirable effects. Once a species has been introduced, it cannot usually be eradicated, although it may be possible to mitigate its undesirable effects.

The difficulty in reversing an introduction and its adverse effects should figure prominently in the decision process on whether to allow an introduction. Assessment of the risks of intentional introductions on fisheries is necessary for the precautionary approach; the ICES Code of Practice provides a procedure for such an assessment.

To encourage more compliance with the precautionary approach to introductions, the existing Code of Practice formulated for the ICES region can be modified and adapted to suit more national implementation of the codes by streamlining its procedures without weakening the rigour of the codes.

Unintended introductions are inherently unprecautionary because they can rarely be evaluated in advance. A precautionary approach would aim at reducing the risk of such unintended introductions and minimize their impact.

Main Issues and Objectives: Introductions are considered here from the perspective of the fishery sector. The main reasons for deliberate introductions include production of protein, employment, generation of foreign exchange, biological control and recreation. Species have been introduced through activities associated with transport (*i.e.*, ballast water, ship and oil-rig fouling), trade in living organisms including aquarium species, aquaculture and fisheries (commercial, recreational, stock enhancement, organisms carried on fishing equipment, live bait fish). Many of these activities have increased over the last century and are expected to increase further in the future.

Potential impacts of some introductions on the fishery sector included changes in the distribution and abundance of fishery resources through disease, changes in predator-prey relationships, changes in competition, mixing of bad (maladapted) genes, and habitat modification. There may also be second and third order changes that affect the ecosystem. Changes in fishing strategy and the fishing community may also require changes in fishing gear and season to allow for a newly introduced species to establish itself or to avoid potential side issues associated with the new fishery. Climate changes may also have significant consequences that may modify the environment, making it more suitable for introductions of either useful or harmful species.

The use of introduced species, including genetically modified and genetically selected organisms, may allow for continued or increased production from habitats that have been so altered or degraded that native fisheries are no longer viable. Care should be taken not to use this potential productivity from introduced species as justification for further abuse of habitat or for delaying their restoration.

Research and Technology

Deliberate Introductions: The ICES Code of Practice (Annex A) describes the research activities that should be conducted in advance of an introduction as follows: (1) desk assessment of the biology and ecology of the intended introduction; (2) preparation of a hazard assessment (detailed analysis of potential environmental impacts); (3) examination of the species within its home range. The results of the above research should be contained in a prospectus or proposal to be submitted to the competent authority for evaluation and decision.

Technological intervention can be used during and following the introduction; such technologies may include:

a. use of hatcheries and quarantine stations to reduce the chance of spread of disease to fishery resources, and to impart some control on numbers of exotic organisms released;
b. use of sterile organisms to reduce the chance of interbreeding with natural fish stocks;
c. genetic stock identification to reduce or prevent genetic changes in the fishery resource;

d. disease diagnostics to monitor the health of the introduced species, and

e. development of the use of limited (pilot) scale introductions to assess impacts and performance.

Continued research and technological intervention on introduced species as part of a monitoring and evaluation scheme should be conducted to assess their impact, health and performance within their new habitat. In this regard, databases or registries of introductions of aquatic species, including their ecological and socio-economic impact, have been established and should be maintained by competent organizations with input from the fishery sector.

Unintended Introductions

Although unintended introductions may arise from several sources, such as fouling organisms, removal of natural barriers and aquarium fish trade, ballast water is probably the most significant and troublesome for the fishery sector and, therefore, emphasized here. In the case of ballast water and sediment, desk studies may be undertaken to determine (1) main ballast water sources, (2) volumes of ballast introduced, and (3) likely "hot spots" as sources of introductions.

Active research should take place and continue on:

a. practical methods for treating organisms in ballast water and sediment;

b. study of dynamics of target species in voyage;

c. study of algal cysts in ships ballast sediment and in port areas;

d. effectiveness of reballasting activities;

e. design changes to ballast water tanks to kill or control harmful species in ballast water, and

f. vessel design that will facilitate the treatment and handling of ballast sediments and water.

Research into new effective non-biocidal antifouling materials should continue to reduce the risks of introductions on ships* hulls and to replace those biocidal applications detrimental for capture fisheries. Antifouling agents are designed to reduce drag and increase the fuel efficiency of a vessel and its long-term efficacy, but they should also be environmentally friendly. Consideration should also be given to those antifouling agents designed to control organisms that would be especially harmful to capture fisheries, even though they may not affect the performance of the vessel.

Management

The first step to managing introductions is to establish a management authority with responsibility for evaluation of proposed introductions, approval in accordance with these guidelines and assuring monitoring of the effects of the introduction. The management of species introductions will involve comparative risk assessment and choices between various options to increase productivity. Management options are limited here to those in the aquatic sector, although countries may be aware of broader issues, such as the development of other sectors (*e.g.*, agriculture). International codes of practice, such as the ICES code, provide a good framework for the management of purposeful species introductions. These codes should not be seen as a hindrance to development, but rather as a tool to help importers make good choices with regard to introductions. Implementation of the code should increase the probability of success of an appropriate introduction.

Deliberate Introductions

Intended introductions should be controlled. As a consequence, those making an introduction should follow the ICES or similar code of practice as appropriate and would be expected to demonstrate caution by preparing a proposal covering: (1) the purpose and objectives of the introduction in advance of the introduction, (2) all relevant biological, ecological, and genetic data of the species in the target area likely to be affected, (3) analysis of potential impacts at the introduction site, including potential ecological, genetic and disease impacts and consequences of its spread, and (4) a qualitative and, where possible, a quantitative risk assessment.

If this proposal is approved: (1) a brood stock should be established at a suitable quarantine site; (2) all effluents from facility should be appropriately sterilized; (3) isolated first generation individuals, free of disease, should be released to the wild in small numbers; and (4) studies of the introduction in the new environment should be continued.

A contingency plan should take account of negative effects should they become apparent and warrant intervention.

The code should also cover introductions that are part of current commercial practice (live trade in fish and shellfish) and recommend:

(a) periodic inspection prior to exportation;

(b) regular inspection; and

(c) quarantine and control if appropriate.

The concern expressed in relation to the introduction of species for fisheries, using the ICES Code, should apply to those species under consideration for biological control, which may have implications for capture fisheries. Bio-control programmes should be weighed carefully against other control methodologies, such as physical and chemical techniques or through intensive fishing. It is likely biological control techniques will take some years to evaluate, through field trials. Much may be learned from the studies on biological control in other disciplines, such as entomology.

The proposal submitted by the potential importer of an exotic species and its review by the competent authority serve as precautionary measures to reduce the chance of a harmful introduction. Governments may wish to consider for national legislation that, if these elements of the code are not followed, the importer of an exotic species may be financially responsible and subject to liability, should significant negative effects arise.

Care should be taken to ensure that the introduced population has an adequate genetic resource base, *i.e.*, genetic diversity, low inbreeding, etc. This may reduce the need for additional introductions, which might otherwise be necessary to increase the genetic resource base. In addition, consideration should be given to the use of gametes, *e.g.*, eggs, cryopreserved sperm, as import material instead of whole organisms to reduce the risk of introducing disease or unintended organisms.

Unintended Introductions

Introduction of undesirable species via ballast water poses problems for fisheries worldwide. In addition to unintentional introductions by ballast water, there are many other mechanisms, including fishing and trade in live fish. Fishing can introduce species by transporting live or fresh bait, or biologically contaminated fishing gear between ecosystems. With trade in live organisms (for aquarium or human consumption), there is the risk of escape.

Authorities responsible for regulating fishing and trade in fish products should establish regulations to reduce these risks, commensurate with the severity of potential adverse impacts. However, the national and international competent authorities that deal with ballast issues are seldom also responsible for fishery management matters. Cooperation between these authorities would greatly aid the management of this problem.

In order to reduce the risk of introductions of organisms in ballast water on capture fisheries in or near deballasting areas, the following methods of prevention include, as recommended by the IMO (1994, Annex B): (a) non-release of ballast; (b) ballast water exchange(s) in or near approved areas; (c) preventing or minimizing uptake of contaminated water or sediment (in shallow water, near dredging operations, during algal blooms); (d) special ballasting facilities on shore; (e) education of crews about ballast-water management procedures, and (f) treatments of ballast water, including changes in temperature and salinity and use of biocides (chemicals).

Although the issues of ballast-water transport, fouling organisms and other unintentional introductions may fall outside the mandate of fishery ministries, the fishery sector could contribute to the management of such introductions, which impact upon the industry. This could be accomplished by promoting the establishment and maintenance in the appropriate institution, of an accessible database on ballast or fouling organisms that have a demonstrated impact on fisheries, by promoting a network of experts who would identify problems, assist with species identification, and delimit areas of impact.

The fishery sector may be well placed to detect the spread of harmful ballast/fouling organisms and should, therefore, contribute to such databases and networks once established, and may take a lead in instigating action on environmental management.

Introduced organisms may cause major changes in ecosystems, especially in port and associated, partly enclosed or enclosed areas, such as lagoons. Such introductions may result in changes in the productivity of local harvested species. Monitoring of introduced organisms and fisheries in these areas may provide useful information as a basis for modifying management techniques and policy for harvested resources.

Spices and Condiments

Spices are those plants, the products of which are made use as food adjuncts to add aroma and flavour. Condiments are also spices, products of which are used as food adjuncts to add taste only. Both spices and condiments contain essential oils, which provide the flavour and taste. They also stimulate digestion on account of there carminative properties. They are of little nutritive value. They are used whole, ground, paste or liquid from, mainly for flavouring and seasoning food. Most spices increase the shelf life of food, especially the dry varieties. Some are added to improve texture and some to introduce a palatable colour or odour.

Classification of Spices

There are about 35 spices and condiments which can be broadly classified into 6 groups, based upon the parts of the plants from which they are obtained:

a. Rhizomes and root spices: Ginger, Turmeric, and Garlic.
b. Seed spices: Nutmeg, Coriander, Fennel, Cumin, Fenugreek, Dill, Aniseed, Celery and Bishopweed.
c. Bark spices: Cinnamon
d. Fruit spices: Cardamom, Black Pepper, Vanilla, All spice, Cassia, Tamarind, etc.
e. Leaf spices: Bay leaf, Curry leaf, etc.
f. Flower spices: Clove, Saffron, Asafoetida, etc.

Importance of Spices in India

Most of the spices are native of our country and hence India is known as the Land of Spices.

1. India produces spices on 2.0 million ha with an annual production of about 2.3 million tonnes, contributing nearly 20% of world's production.
2. Nearly 90-95% of the total production is consumed locally and the rest exported.
3. India is the biggest exporter of spices.
4. They are indispensable part of our culinary preparations especially used for flavouring and seasoning of food.
5. Most of the spices have potential medicinal values. Besides, the spices and spice products are also indirectly used as flavouring or colouring agents or as preservatives in many pharmaceutical preparations.
6. Spices have been used in cosmetic and perfumery industries. Spice oils are used in the manufacture of soaps, tooth pastes, talcum powder, aftershave lotions, vanishing creams, mouth freshners and room freshners etc.,

The Indian fish fauna is divided into two classes, *viz.*, Chondrichthyes and Osteichthyes. The Chondrichthyes are represented by 131 species under 67 genera, 28 families and 10 Orders in the Indian region. The annual average landings of the Indian Chondrichthyes is 33,442 tonnes, of which, 15,537 tonnes come from the east coast and 17,605 tonnes come from the west coast and the rest come from the Andaman and Nicobar, and Lakshadeep Islands.

The Indian Osteichthyes are represented by 2,415 species belonging to 902 genera, 226 families and 30 orders, of which, five families, notably the family Parapsilorhynchidae are endemic to India. These small hillstream fishes include a single genus, *viz.*, *Parapsilorhynchus* which contains 3 species. They occur in the Western Ghats, Satpura mountains and the Bailadila range in Madhya Pradesh only. Further, the fishes of the family Psilorhynchidae with the only genus Psilorhynchus are also endemic to the Indian region. Other fishes endemic to

India include the genus *Olytra* and the species *Horaichthys* setnai belonging to the families Olyridae and Horaichthyidae respectively. The latter occur from the Gulf of Kutch to Trivandrum coast. The endemic fish families form 2.21 per cent of the total bony fish families of the Indian region. 223 endemic fish species are found in India, representing 8.75 per cent of the total fish species known from the Indian region and 128 monotypic genera of fishes found in India, representing 13.20 per cent of the genera of fishes known from the Indian region.

About 22000 species of fishes have been recorded in the world; of which, about 11% are found in Indian waters. Out of the 2200 species so far listed, 73 (3.32%) belong to the cold freshwater regime, 544 (24.73%) to the warm fresh waters domain, 143 (6.50%) to the brackish waters and 1440 (65.45%) to the marine ecosystem. Adequate protection of ecosystems is a necessary requirement for survival of all species and proper care is needed to overcome anthropogenic stresses. In the case of commercial species, rational exploitation is a pre- requisite for sustainability of the resources.

There are about 450 families of freshwater fishes globally. Roughly 40 are represented in India (warm freshwater species). About 25 of these families contain commercially important species. Number of endemic species in warm water is about 544. Major warm water species are:

Bagarius bagarius, Catla catla, Channa marulius, C. punctatus, C. striatus, Cirrhinus mrigala, Clarias batrachus, Heteropneustes fossilis, Labeo bata, L. calbasu, L. rohita, Aorichthys seenghala, Notopterus chitala, N. notopterus, Pangasius pangasius, Rita rita, Wallago attu.

Cyprinids (family: Cyprinidae), Live fish (family: Anabantidae, Clariidae, Channidae, Heteropneustidae), Cat fish (family: Bagridae, Silurdae, Schilbeidae), Clupeids (family: Clupeidae), Mullets (family: Mugilidae), featherbacks (family: Notopteridae), Loaches (family: Cobitidae), Eels (family: Mastacembelidae), Glass fishes (family: Chandidae) and Gobies (family: Gobiidae) are the major groups of fresh water fishes found in India.

Cyprinidae is one of the largest families and is well represented in India with species ranging from few milimeters in length (minnows) to more than a metre (major carps). Among the I 544 freshwater fish species in India, Cyprinidae accounts for nearly 24.12% of them.

Many of the Cyprinids, especially carps, are widely captured and form the mainstay of culture operations. Depending on their growth and utility in culture systems, the carps are grouped as the major carps (*Catla catla, Cirrhinus mrigala, Labeo calbasu, Labeo rohita*) and minor carps (*L. fimbriatus, L. bata, C. cirrhosus, C. reba*). A few exotic carps, as indicated below, have also been introduced into the country mainly for culture purposes (*Hypopthalmichthys molitrix, Ctenopharyngodon idella, Cyprinus carpio*). The mahseers are also included in cyprinidae. Though many species of the genus *Tor* are found in the high altitude cold waters, some species like *Tor khudree* and *Tor putitora* occur in warmer waters. The genus *Tor* thus reveals a wide range of species diversity as regards adaptation to different ecological conditions. There is a serious decline of mahseer fishery in different ecosystems endangering its very existence, thus warranting adequate conservation of the fishes.

Objectives of this study are to (a) focus on the diversity of fish species population and their conservation in an aquatic ecosystem and suggest conservation measures to preserve bio-diversity and (b) to explore the relationships between chemical complexes of the lake water, soil and fish yield from a lake. The Sone lake located in Assam is chosen to carry out these investigations.

The water volume of lakes and rivers total around 126,000 km3, which is equivalent to 0.0093% of the total volume of liquid water in the world (WCMC, 1992). The province of Assam (in India) is gifted with myriads of riverine and tectonic Lakes (locally called "Beels") which alone constitute 81% (approx.) of the total lentic area (> 1,20,000 ha) in Assam. In Assam, there are 1392 (approx.) Beels with a water spread area of 50,000 ha (approx.), of which, 50% of the Beel area has been derelicted. Nevertheless, contrary to average annual fish production of 6-7 kg per hectare from Indian reservoirs, the overall fish production from the beels of Assam is more than 100 Kg per hectare per year (Dey, 1981).

Seventy ichthyospecies belonging to 49 genera under 24 families have been recorded in lake Sone. Of these, 84.2% belong to the primary freshwater group, while the rest to the peripheral class. On the basis of Indian and Extra-Indian territorial distribution, the fish species of Sone include 20 widely distributed species and eight species of Northern India. Further, ecomorphologically, the true rheophilic group contains only one species while the semi-torrential group includes five species. The migratory class contains seven species and the plainwater group includes 57 fish species of the Lake.

Sone is the largest freshwater tectonic lake of Assam with an area of 3458.12 hectares (ha) at full storage level (FSL), 409.37 ha at dead storage level (DSL), maximum length 12.5 km, maximum breadth 3.9 km and shoreline 35.4 km. Sone is a syncline near the Badarpur line of folding and the Chargola anticline. It belongs to the Mio-pliocene epoch with alluvium in the Lake bed and fine grain sand stones in the catchment. The catchment also has evergreen forest. This Lake is continuously fed by the major inlet Singla which originates as Thing Tlawng Lui from the Mizo Hills (approximately 365 m MSL) and after traversing a tortuous course of approx. 62 km, it enters Lake Sone. The major outlet, river Kachua, drains the lake water into the river Kushiyara after traversing approx. 19 km from the northernmost end of the Lake.

Monthly statistics of fish yield, water and soil samples were collected for two years. The data were analysed for computation of diversity indices, fish population dynamics and impact of physico chemical complexes on fish yield of the lake.

Diversity encompasses two different concepts of variety and variability; *viz.*, richness and evenness. These two concepts, in theory, can be applied across a hierarchy of scales, from genetic diversity through to ecosystem diversity (Burton, et al., 1992). In the present study, diversity was measured by the number of species (species richness) and by using the two indices namely Shannon-Weaver [H'] and Simpson's [D] Indices. These are given by -

a. Shannon-Weaver index of diversity (Shannon, and Weaver, 1949; Wolda, H. 1983)

$H' = - pi\ 1n\ pi$

where pi is the proportion of the ith species in the sample. The samples were collected from a large fish community, in which, the total number of species is known.

b. Simpson's index of diversity (Simpson, 1949)

$D = 1 - (p1^2)$

Physico-Chemical Analyses: The physical and chemical parameters of the water and soil have been analysed respectively through water and soil sampling at fortnightly intervals. During 24 months field investigations, thirteen independent variables were estimated: water temperature (WT), turbidity (t), water pH (WpH), dissolved oxygen (DO), free carbon dioxide (FCO_2), total alkalinity (TA), water conductivity (WC), soil temperature (ST), soil pH (SpH), soil organic carbon (SOC), soil phosphorus (SP), soil potassium (SK) and aquatic macrophytic biomass (AMB). In order to look at the relationship between these variables and fish yield, regression/correlation analyses were carried out. Twenty five different equations (both linear and quadratic relationships) were looked at, and the best relationship is selected based on strong correlation coefficient and least percentage of standard error of Y estimate (dependent variable). Finally, the step-wise forward selection method was used in a linear multiple regression of fish yield (FY) against the independent variables.

The stepwise procedure made no allowance for the possibility that a variable entered as significant might become insignificant on the addition of further variables into equation, and a variable entered into regression equation and subsequently removed (after addition of several other variables) would indicate that the variable is significant in reducing the variation. The final regression thus indicates only significant ($p <= 0.05$) variables.

Data collected for twenty four months in lake Sone is analysed for computation of diversity indices and fish population dynamics and analyses of limnological parameters of the Lake.

The fish community of lake Sone is diverse with 70 species of different size classes. Diversity of fish species is also reflected in the different types of zoogeographical groups. Population dynamics of the Indian Major Carps (IMC), clupeids, silurids and bagrids have been dealt in this paper due to their commercial importance and significant contribution to overall fish production of the lake.

The population of IMC, as a group, fluctuated with a single peak in December during the first session and two peaks, one each in September and January, during the second session. The rich landing during the dry season is essentially due to high catch at low water level (WL) and the monsoon influx could be due to more IMC migration into the Lake from the adjoining rivers.

Cirrhinus mrigala portrays two peaks, one each in August and January. Labeo calbasu has rich population in September and little higher population in January. L. gonius exhibits peaks in July and in September. L. rohita depicts peaks in August and in January. P. chola, among the minor carps, is a dominant species in the lake contributing significantly to the total fish production.

The Clupeids of the Lake consists of Gudusia chapra and Hilsa ilisha. Population dynamics of the group behaved uniformly during both the sessions. The seasonal fluctuation of the group is essentially due to G. chapra with peaks in August and in December. H. ilisha is an anadromous migrant into Sone ecosystem in monsoon season only unlike two (monsoon and winter) in other freshwater bodies.

The siluroids show uniform trend in their dynamics. Among the two species in this group, Ompok bimaculatus displays single peak in January while, Wallago attu has maximum population in August and in January.

The bagrids, as a group, shows peaks in October and in February in the first session and, August and November in the second session. Among the bagrids, Mystus vittatus is the most dominant fish which had accounted almost entirely for the dynamics of the whole group and depicted a peak each in July-August and November. Aorichthys seenghala has little high population trend during April, June, January and February months.

Both H' and D follow same pattern of fluctuation with peaks in monsoon season. This lake is highly diverse during monsoon in both sessions with a rising trend beginning from July. It is seen that, species like P. chola, G. chapra, O. bimaculatus and M. vittatus contributed significantly for the high production in monsoon season. During these two sessions, there had been no sign of decreasing trend in diversity under stressed condition However, the decreasing trend in diversity indices during other seasons were mainly due to competition and intense fishing.

Before 1950, lake Sone had not been subjected to much man-made disturbances and it exhibited high species diversity in a stable environment; and, possibly because of this, there had not been evolution of finer specialisation and adaptations which resulted in smaller niches and more species occupying a unit space of habitat. It could also be argued that the high diversity of lake Sone fish communities has not been achieved only by precise niche' diversification in an equilibrium community but rather by highly variable larval recruitment which causes a competition for vacant living spaces in which the first to arrive wins. Since 1950, after the influx of large number of fishermen from adjoining country (Bangladesh) to lake Sone, the biotope had turned into a disturbed environment due to tremendous fishing and agricultural activities, which have been exerting substantial influence on the fish population of the Lake. In view of these, Sone ecosystem of today could be regarded as a disturbed ecosystem but, at the same time, it has been preserving its rich fish diversity as compared to many undisturbed ecosystems where competition and exclusion are more intense.

It may be noted here that the introduction of common carp (Cyprinus carpio) into Sone, about two decades ago, may have significant impact on the diversity of indigenous fish species in future. Aim of this exercise was to transform slow growing fisheries into a rapid one and also to compensate for the loss caused to the fishery of the Lake by the construction of a blind dam and then its replacement by a lock gate (without fish passes and fish ladders) over Kachna (the major outlet) during 1951 and 1964 respectively.

These hindrances have reduced the fresh water pulse and might have triggered the deviation of the migratory path of the fishes and take a lengthier route in order to enter Lake Sone. However, a serious effect of this single species introduction leads to inability to maintain genetic diversity inherent in naturally reproducing populations.

The limnological parameters did not show much abrupt fluctuation with WT ranging from 18.7 to 32.3 deg C, turbidity 20.56 to 185.54 TU, WpH 6.0 to 7.9, DO 2.43 to 5.96 mg/lit, FCO_2 0.90 to 14.55 mg/lit, TA 25.0 to 76.0 mg/lit, WC 0.000051 to 0.000039 mhos/cm; ST 19.9 to 32.3 deg C, SpH 5.09 to 5.99, SOC 0.25 to 1.74%, AP 0.15 to 1.93 mg/100g and AK 1.6 to 24.8 mg/100g. The AMB varied from 0.50 to 21.90 kg per sq. metre.

The average fish yield from Lake Sone has been estimated to be 96.8 kg per ha. Species such as Puntius chola contributed to bulk of the Lake harvest registering an annual relative yield (ARY) of 24.60%. The IMC depicted an average ARY of 0.73%, in which, Labeo rohita is 0.24%. Of all the predators, with ARY lying in between 1.75 and 4.83%, Wallago attu alone contributed to an average ARY of 2.94%. Hilsa ilisha, with an ARY of 0.04%, revealed a single run during the monsoon against two in other Indian waterbodies.

Regression analyses of the variables SOC and FY, AMB and FY showed linear relationships as indicated below:

a. fish yield FY = 49146.34 - 22654.3 (SOC), R=0.60 (standard error of Y estimate = 11.25%)
b. fish yield FY = 32600.18 - 836.04 (AMB), R=0.39 (standard error of Y estimate = 31.02%).

Analysis of the correlation between FY and some of the limnological parameters, as referred to above, were also done through quadratic equations. This reveals significant relations of FY with the parameters WpH, TA, WC, SOC, SK and AMB (based on respective correlation coefficient values). These

quadratic relationships between FY and parameters, based on their respective best equations, are given below:-

i. **FY vs WpH:**

a. $FY = 1/(0.11940 * (WpH + (-6.996)^2 + 0.02812)$, $R = 0.486$

b. $FY = 31.550 * e^\wedge ((\ln(pH) - 1.931)^2/(-0.01754))$, $R = 0.470$.

ii. **FY vs TA**

a. $Y = 31.37 * e^\wedge ((\ln(TA) - 3.922)2/(-0.2397))$, $R = 0.469$.

This result, shows maximum FY at a TA value of 50 and gradual decreasing yield trend at TA values both <50 and >50, corroborates with the field condition in the studied Lake.

iii. **FY vs WC: This relationship is found to be power law and is given by**

$Y = 40.90 * 0.9935^\wedge WC$, $R = 0.668$.

indicates gradual decreasing trend in FY at high values of WC- a situation which reflects the field condition in the studied biotope.

iv. FY vs SOC

$Y = 23.06 * SOC^\wedge (-1.313 * SOC)$, $R = 0.713$.

reflected the actual condition in the field with a gradual fall of FY at higher SOC values because more SOC reveals more organic decomposition.

v. FY vs SK

a. $Y = [\ 17.88 + 3.2 * SK + (-0.162) * SK * SK\ ]$, $R = 0.459$.

b. $Y = 32.69 * e^\wedge ((SK-10.60)/2)$, $R = 0.5656$.

portrayed a gradual decreasing trend of FY from SK values > 10 mg/100g upto which there has been an increasing trend from an initial high value.

vi. FY vs AMB

$Y = 30.56 * 0.9387^\wedge AMB$ $R = 0.578$.

$Y = 1/(-0.0004475 + (AMB) + (-5.946))2 + (-0.4520)$, $R = 0.848$.

revealed a decreasing trend of FY with higher AMB in the Lake. This tallies with field situation because more AMB would make less space available as fish habitat besides hindering their free movement.

The stepwise multiple regression of limnological parameters indicate that the variartions in fish yield are mostly caused by DO, WpH, t, TA and ST. while variation in fish production due to variables such as FCO_2, WC, SpH, SOC, SP, SK and AMB are less. These analyses clearly indicate that factors such as DO, WpH, t, TA, ST and to some extent WT determines the fish yield in Lake Sone eco-system.

Conservation programmes help fish production to be more sustainable while at the same time maintaining diversity. Conserving diversity also improves the likelihood of maintaining minimal viable populations of rare and late-successional species. Maintaining ichthyo diversity is important because it is not always possible to identify which individual species are critical to aquatic ecosystem's sustainability. Many fish species may provide genetic material and may serve as ecological indicators. Diversity reduces disease problems and encourages recovery from disturbance. Although, over the past billion years, adaptation and diversification has tended to increase the number of species, with the escalation of human population, their infringing into the wilderness and rapid industrialisation, a decline in species diversity became noticeable. This decline is sometimes associated with the destruction of aquatic ecosystems. The situation is rather in an aggravated condition in the Lakes like Sone ecosystem situated in the tropical developing countries where the process of species extinction and genetic loss may become severe in future due to loss of habitat by siltation, blockade of the waterways, injudicious introduction of exotic fish species, irrational harvest of the juveniles and fish diseases (Kar, et.al. 1995b).

Fish conservation in Sone may be practiced either in situ or ex situ. Methods such as secluding portions of the Lake as 'aquareserves' with minimum viable population size in it could be practiced as in situ conservation strategy in the lake Sone. Ex situ conservation may involve preservation of the gene pool as an insurance against loss in situ reserves. In situ is best achieved both in natural and man-made ecosystems. The concept mainly revolves around the conservation of endemic species. This method has the advantage of being less expensive than ex situ measures in the long run and also allows continued co-evolution with other species. Ex situ approach takes care of species outside their natural habitats, especially in farms or genetic resource conservation centres. The ultimate ex-situ measures are the gene banks where cryostorage of gametes and embryos are achieved.

A number of stresses, mostly man made, have been operating on fishes, to the grave detriment of these species in their once ideal environments. The obvious fallout of these on commercial species is through reduced catch per unit effort. Unlike commercial species, where depletion or decline of numbers is easily detectable, and some remedial measures taken, uneconomical species, though a part of the bio-diversity, may not even get noticed even if affected.

Reclamation of water bodies for construction of buildings and agricultural purposes destroys the natural habitat of fishes. Due to the increasing human population pressure, many of the natural water bodies have either reduced in size or completely disappeared. Removal of bottom sand and gravel from river beds has been another menace destructing the natural spawning grounds of fishes. Bottom feeders are also affected as the bottom flora and fauna get removed alongwith the gravel collections.

Over-exploitation had been having tremendous bad impact on fish diversity. Notopterus chitala and Semiplotus semiplotus, which used to be abundant in the large rivers of Assam and attaining large sizes, are becoming rarer now due to over exploitation. Unconventional and illegal fishing methods, like use of banned nets, electric fishing, use of explosives, etc. cause mass mortality, and kill the unwanted fishes in addition to the commercial species. Many of the smaller fishes are also killed using these methods. In the case of mahseer, the main causes of depletion are wanton destruction of brood fish and juveniles and the deteriorating ecological conditions of the spawning and rearing grounds. Large sized mahseers are more convenient targets. Discarded food articles are used as bait and dynamite is thrown when large and small fish congregate. This results in wholesale slaughter of the entire population.

Introduction of exotic species as a part of aquaculture for commercial gains has resulted in loss of diversity. For *e.g.* the Schizothoracine fishes in Kashmir valley have almost been exterminated by the exotic common carp Cyprinus carpio species. The Loktak fish of Manipur is fast disappearing, once again, due to the introduction of common carp for culture. The displacement of Catla species from many reservoirs and its associated ecological disaster is noticed after introduction of silver carp as in Govindsagar and some other reservoirs. The consequences of pollution in water bodies due to industrial and domestic pollutants are far reaching. The obvious outcome of pollution is stress and death of individuals. In cases of genotoxic pollutants, the effect can be more damaging but subtle. A number of fishes have been displaced or eliminated from their original habitats. Among the range of various diseases caused by bacteria, fungi, virus, etc. the most virulent and menacing one threatening many species is the Epizootic Ulcerative Syndrome (EUS). It has wiped out large population of a number of commercial and non-commercial species in Asian region resulting in severe erosion of genetic variability and bio-diversity.

Though a number of causes have been identified, it has been not possible to totally undo the damage caused by each factor. A holistic approach integrating the concept of sustainable development and conservation measures can help. Control measures such as, diversion of effluent (for treatment) in some lakes have shown appreciable improvement in the levels of a number of ecological parameters like dissolved oxygen, free carbon dioxide total alkalinity etc. Fish production potentials improved from 5-39% in just one year.

A typical fishery in tropical waters may harvest a hundred different species, each species often consisting of several local groups with a distinct genetic make-up. A detailed study of each group is difficult. There will be uncertainties with all scientific endeavours to monitor abundance and productivity of stocks and the underlying causes. Further, there is an uncertainty with regard to climate, aquatic ecosystem productivity, predation and fishing pressure. Managing for ichthyo diversity provides a useful clue against these uncertainties. Managing for fish diversity represents the epitome of water stewardship, quite distinct from simply maintaining reliable stocks of individual species. Fishers and ichthyologists have a critical role to play in understanding and protecting diverse fish resources. The long-term ability of this aquatic regime to sustain the generation of fish and fish products and values that we expect of them may depend on this ichthyo species diversity.

Concomitant to the above, the ongoing process of ecosystem change, as is evident in lake Sone today, directly or indirectly affects the abundance and composition of the fish species; and, is, to a large and increasing degree, due to human activities which are sometimes unconnected to fisheries. Such increasing impacts of other human activities on the lake Sone ecosystem highlights the need for a regional aquatic ecosystem approach to fisheries management including water-use patterns. The state of the fish community may be seen as a valid integrative indicator of aquatic ecosystem quality and health and - somewhat more distantly - of the regional quality of life for the humans.

We find that lake Sone is a vast aquaregime with rich ichthyo diversity of 70 species, the diversity followed same pattern during both the sessions. The conservation of fish diversity assumes topmost priority under changing circumstances of gradual habitat degradation. Whether or not the Sone aquatic ecosystem would be able to maintain its ichthyo diversity would be largely determined by the water-use patterns since fish diversity is an interdependent part of a cluster of diversities in the human society and in the aquatic ecosystem. In order to assure ecological and economic sustainability of the lake Sone fisheries, matching human activities with the natural pulses of the ecosystem should be particularly stressed in management practices rather than basing such practice on an unrealistic expectation that fish populations provide a steady exploitation source.

The average fish yield from Sone has been estimated to be 96.8 kg per hectare. Species such as Puntius chola contributed to bulk of the lake fish landing registering an annual relative yield of 24.60%. This study reveals that in Lake Sone, (a) maximum fish yield occurs in pH range of 6.9 - 7.1. (b) fish yield declines with increase in WC, SOC and AMB. The stepwise regression analyses indicate that the variartions in fish yield are mostly caused by DO, WpH, t, TA and ST. while variation in fish production due to variables such as FCO_2, WC, SpH, SOC, SP, SK and AMB are less. These analyses clearly indicate that factors such as DO, WpH, t, TA, ST and to some extent WT determines the fish production in lake Sone eco system.

This work was supported partly by the grants from the Ministry of Environment and Forests, Government of India and Indian Institute of Science. We thank Prof. Madhav Gadgil and Prof. N.V.Joshi for support and fruitful exchange of views. We thank Indian Institute of Science, Jawaharlal Nehru Centre for Advanced Scientific Research and Assam University for extending cooperation to carry out this research. We wish to acknowledge the help received from the fisherfolk and colleagues during the field work and analyses.

Extinction Event

An extinction event (also known as: mass extinction; extinction-level event, ELE) occurs when there is a sharp decrease in the number of species in a relatively short period of time. Mass extinctions affect most major taxonomic groups present at the time — birds, mammals, reptiles, amphibians, fish, invertebrates and other simpler life forms. They may be caused by one or both of:

- extinction of an unusually large number of species in a short period.
- a sharp drop in the rate of speciation.

Based on the fossil record, the background rate of extinctions on Earth is about two to five taxonomic families of marine invertebrates and vertebrates every million years. Marine fossils are generally used to measure extinction rates because they are more plentiful and cover a longer time span than fossils of land organisms. Apparent extinction intensity, *i.e.* the fraction of genera going extinct at any given time, as reconstructed from the fossil record. (Graph not meant to include recent epoch of Holocene extinction event)

Since life began on Earth, several major mass extinctions have significantly exceeded the background extinction rate. Although there probably were mass extinctions in the Archean and Proterozoic, it is only during the Phanerozoic Eon that the evolution of organisms with hard body parts has provided a sufficient fossil record from which to make a systematic study of extinction patterns.

Estimates of the number of major mass extinctions in the last 540 million years range from as few as five to more than twenty. These differences stem primarily from: (a) the threshold chosen for describing an extinction event as "major"; (b) what set of data one chooses as the measure of past diversity.

Major Extinction Events

The classical "Big Five" mass extinctions identified by Raup and Sepkoski (1982) are widely agreed upon as some of the most significant: End Ordovician, Late Devonian, End Permian, End Triassic, and End Cretaceous.

These and a selection of other extinction events are outlined below. The articles about individual mass extinctions describe their effects in more detail and discuss theories about their causes.

1. 488 million years ago—a series of mass extinctions at the Cambrian-Ordovician transition (the Cambrian-Ordovician extinction events) eliminated many brachiopods and conodonts and severely reduced the number of trilobite species.
2. 444 million years ago—at the Ordovician-Silurian transition two Ordovician-Silurian extinction events occurred, and together these are ranked by many scientists as the second largest of the five major extinctions in Earth's history in terms of percentage of genera that went extinct.

3. 360 million years ago — near the Devonian-Carboniferous transition (the Late Devonian extinction) a prolonged series of extinctions led to the elimination of about 70% of all species. This was not a sudden event — the period of decline lasted perhaps as long as 20 million years, and there is evidence for a series of extinction pulses within this period.
4. 251 million years ago — at the Permian-Triassic transition Earth's largest extinction (the P/Tr or Permian-Triassic extinction event) killed 53% of marine families, 84% of marine genera, about 96% of all marine species and an estimated 70% of land species (including plants, insects, and vertebrate animals). The "Great Dying" had enormous evolutionary significance: on land it ended the dominance of the mammal-like reptiles and created the opportunity for archosaurs and then dinosaurs to become the dominant land vertebrates; in the seas the percentage of animals that were sessile dropped from 67% to 50%.

   The whole of the late Permian was a difficult time for at least marine life - even before the "Great Dying", the diagram shows a late-Permian level of extinction large enough to qualify for inclusion in the "Big Five".
5. 200 million years ago — at the Triassic-Jurassic transition (the Triassic-Jurassic extinction event) about 20% of all marine families as well as most non-dinosaurian archosaurs, most therapsids, and the last of the large amphibians were eliminated.
6. 65 million years ago — at the Cretaceous-Paleogene transition (the K/T or Cretaceous-Tertiary extinction event) about 50% of all species became extinct. It has great significance for humans because it ended the reign of the dinosaurs and opened the way for mammals to become the dominant land vertebrates; and in the seas it reduced the percentage of sessile animals again, to about 33%. The K/T extinction was rather uneven— some groups of organisms became extinct, some suffered heavy losses and some appear to have got off relatively lightly.

7. Present day — the Holocene extinction event. A 1998 survey by the American Museum of Natural History found that 70% of biologists view the present era as part of a mass extinction event, possibly one of the fastest ever. Some, such as E. O. Wilson of Harvard University, predict that man's destruction of the biosphere could cause the extinction of one-half of all species in the next 100 years. Research and conservation efforts, such as the IUCN's annual "Red List" of threatened species, all point to an ongoing period of enhanced extinction, though some offer much lower rates and hence longer time scales before the onset of catastrophic damage. The extinction of many megafauna near the end of the most recent ice age is also sometimes considered a part of the Holocene extinction event. Some paleontologists, however, question whether the available data support a comparison with mass extinctions in the past.

Evolutionary Importance of Mass Extinctions

Mass extinctions have sometimes accelerated the evolution of life on earth. When dominance of particular ecological niches passes from one group of organisms to another, it is rarely because the new dominant group is "superior" to the old and usually because an extinction event eliminates the old dominant group and makes way for the new one.

For example mammaliformes ("almost mammals") and then mammals existed throughout the reign of the dinosaurs, but could not compete for the large terrestrial vertebrate niches which dinosaurs monopolized. The end-Cretaceous mass extinction removed the non-avian dinosaurs and made it possible for mammals to expand into the large terrestrial vertebrate niches.

On the other hand many groups which survive mass extinctions do not recover in numbers or diversity, and many of these go into long-term decline. This phenomenon is known as "Dead Clade Walking". So most "what died and what survived" analyses understate the full effect of mass extinctions.

Is the frequency of extinctions decreasing?

The appears to show that:

- The gaps between mass extinctions are becoming longer.
- The average and background rates of extinction are decreasing.

The idea that mass extinctions are becoming less frequent is rather speculative - from a statistical point of view a sample of about 10 extinction events is too small to be a reliable sign of any actual trend. But the average and background rates of extinction are based on hundreds of samples over a period of 550M years, so the apparent decrease in these rates is statistically significant and needs to be explained.

Both of these phenomena could be explained by one or more of:

- Reasonably complete fossils are very rare, most extinct organisms are represented only by partial fossils, and complete fossils are rarest in the oldest rocks. So paleontologists have mistakenly assigned parts of the same organism to different genera which were often defined solely to accommodate these finds (the story of *Anomalocaris* is a good example). The risk of this mistake is higher for older fossils because these are often unlike parts of any living organism. Many of the "superfluous" genera are represented by fragments which are not found again and the "superfluous" genera appear to become extinct very quickly.
- Martin (1994, 1996) has argued that the oceans have become more hospitable to life over the last 500M years and less vulnerable to mass extinctions: dissolved oxygen became more widespread and penetrated to greater depths; the development of life on land reduced the run-off of nutrients and hence the risk of eutrophication and anoxic events; and marine ecosystems became more diversified so that food chains were less likely to be disrupted.

Causes of Mass Extinction

There is still debate about the causes of all mass extinctions before the Holocene.

Looking for the Causes of Particular Mass Extinction

Should: (i) explain all of the losses, not just focus on a few groups (such as dinosaurs); (ii) explain why particular groups of organisms died out and why others survived; (iii) provide killing mechanisms which are strong enough to cause a mass extinction but not a total extinction; (iv) be based on events or processes that can be shown to have happened, not just inferred from the extinction.

It may be necessary to consider combinations of causes. For example the marine aspect of the end-Cretaceous extinction appears to have been caused by several processes which partially overlapped in time and may have had different levels of significance in different parts of the world.

Arens and West (2006) proposed a "press/pulse" model in which mass extinctions generally require 2 types of cause: long-term pressure on the eco-system ("press") and a sudden catastrophe ("pulse") towards the end of the period of pressure.

Their statistical analysis of marine extinction rates throughout the Phanerozoic suggested that neither long-term pressure alone nor a catastrophe alone was sufficient to cause a significant increase in the extinction rate.

Most Widely-supported Explanations

Macleod (2001) summarised the relationship between mass extinctions and events which are most often cited as causes of mass extinctions, using data from Courtillot *et al* (1996), Hallam (1992) and Grieve *et al* (1996):

- Flood basalt events: 11 occurrences, all associated with significant extinctions.
- Sea-level falls: 12, of which 7 were associated with significant extinctions.
- Asteroid impacts producing craters over 100km wide: 1, associated with 1 mass extinction.
- Asteroid impacts producing craters less than 100km wide: over 50, the great majority not associated with significant extinctions.

The most commonly-suggested causes of mass extinctions are:

1. ***Flood Basalt Events:*** **The formation of large igneous provinces by flood basalt events could have: produced dust and particulate aerosols which inhibited photosynthesis and thus caused food chains to collapse both on land and at sea; emitted sulphur oxides which were precipitated as acid rain and poisoned many organisms, contributing further to the collapse of food chains; emitted carbon dioxide and thus caused sustained global warming once the dust and particulate aerosols dissipated. Flood basalt events occur as pulses of activity punctuated by dormant periods. As a result they are likely to cause the climate to oscillate between cooling and warming, but with an overall trend towards warming as the carbon dioxide they emit can stay in the atmosphere for hundreds of years.**

   Various scientists have suggested that massive volcanism caused or contributed to the End-Cretaceous, End-Permian, End Triassic and End Jurassic extinctions.

2. ***Sea-level Falls:*** **These are often clearly marked by worldwide sequences of contemporaneous sediments which show all or part of a transition from sea-bed to tidal zone to beach to dry land - and where there is no evidence that the rocks in the relevant areas were raised by geological processes such as orogeny. Sea-level falls could reduce the continental shelf area (the most productive part of the oceans) sufficiently to cause a marine mass extinction, and could disrupt weather patterns enough to cause extinctions on land.**

   But sea-level falls are very probably the result of other events, such as sustained global cooling or the sinking of the mid-ocean ridges.

   Sea-level falls are associated with most of the mass extinctions, including all of the "Big Five"—End-Ordovician, Late Devonian, End-Permian, End-Triassic, and End-Cretaceous.

3. *Impact Events:* **The impact of a sufficiently large asteroid or comet could have caused food chains to collapse both on land and at sea by producing dust and particulate aerosols and thus inhibiting photosynthesis. If it hit sulphur-rich rocks, it could also have emitted sulphur oxides which were precipitated as acid rain and poisoned many organisms — contributing further to the collapse of food chains. Some scientists have suggested that impacts could also have caused megatsunamis and/or global forest fires, but these ideas are now regarded as exaggerations.**

   Only the end-Cretaceous extinction is associated with strong evidence of such an impact, but that impact is easily the largest for which there is strong evidence.

4. *Sustained Global Cooling:* This could: kill many polar and temperate species and force others to migrate towards the equator; reduce the area available for tropical species; often make the Earth's climate more arid on average, mainly by locking up more of the planet's water in ice and snow. The glaciation cycles of the current ice age are believed to have had only a very mild impact on biodiversity, so the mere existence of a significant cooling is not sufficient on its own to explain a mass extinction.

   It has been suggested that global cooling caused or contributed to the End-Ordovician, Permian-Triassic, Late Devonian extinctions, and possibly others.

   Note: this item is labelled "sustained global cooling" to distinguish it from the temporary climatic effects of flood basalt events or impacts.

5. *Sustained Global Warming:* This would have the opposite effects: expand the area available for tropical species; kill temperate species or force them to migrate towards the Poles (or perish); possibly cause severe extinctions of polar species; often make the Earth's climate wetter on average, mainly by melting ice and snow and thus increasing the volume of the water cycle. It might also cause anoxic events in the oceans.

The most dramatic example of sustained warming is the Paleocene-Eocene Thermal Maximum, which was associated with one of the smaller mass extinctions.

6. *Clathrate Gun Hypothesis:* Clathrates are composites in which a lattice of one substance forms a cage round another. Methane clathrates (in which water molecules are the cage) form on continental shelves. These clathrates are likely to break up rapidly and release the methane if the temperature rises quickly or the pressure on them drops quickly - for example in response to sudden global warming or a sudden drop in sea level or even earthquakes. Methane is a much more powerful greenhouse gas than carbon dioxide, so a methane eruption ("clathrate gun") could cause rapid global warming or make it much more severe if the eruption was itself caused by global warming.

   The most likely signature of such a methane eruption would be a sudden decrease in the ratio of carbon-13 to carbon-12 in sediments, since methane clathrates are low in carbon-13; but the change would have to be very large, as other events can also reduce the percentage of carbon-13.

   It has been suggested that "clathrate gun" methane eruptions were involved in the end-Permian extinction and in the Paleocene-Eocene Thermal Maximum, which was associated with one of the smaller mass extinctions.

7. *Anoxic Events:* These are situations in which the upper and even the middle layers of the ocean become deficient or totally lacking in oxygen. Their causes are complex and controversial, but all known instances are associated with severe and sustained global warming, mostly caused by massive sustained volcanism.

   It has been suggested that anoxic events caused or contributed to the late Devonian, Permian-Triassic and Triassic-Jurassic extinctions. On the other hand there are widespread black shale beds from the mid-Cretaceous which indicate anoxic events but are not associated with mass extinctions.

8. *Hydrogen Sulfide Emissions from the Seas:* Kump, Pavlov and Arthur (2005) have proposed that during the Permian-Triassic extinction the warming also upset the oceanic balance between photosynthesising plankton and deep-water sulfate-reducing bacteria, causing massive emissions of hydrogen sulfide which poisoned life on both land and sea and severely weakened the ozone layer, exposing much of the life that still remained to fatal levels of UV radiation.
9. *Oceanic Overturn:* This is a disruption of thermo-haline circulation which lets surface water (which is more saline than deep water because of evaporation) sink straight down, bringing anoxic deep water to the surface and therefore killing most of the oxygen-breathing organisms which inhabit the surface and middle depths. It may occur either at the beginning or the end of a glaciation, although an overturn at the start of a glaciation is more dangerous because the preceding warm period will have created a larger volume of anoxic water.

   Unlike other oceanic catastrophes such as regressions (sea-level falls) and anoxic events, overturns do not leave easily-identified "signatures" in rocks and are theoretical consequences of researchers' conclusions about other climatic and marine events.

   It has been suggested that oceanic overturn caused or contributed to the late Devonian and Permian-Triassic extinctions.
10. *A Nearby Nova, Supernova or Gamma ray Burst:* A nearby gamma ray burst (less than 6000 light years distance) could sufficiently irradiate the surface of the Earth to kill organisms living there and destroy the ozone layer in the process. From statistical arguments, approximately 1 gamma ray burst would be expected to occur in close proximity to Earth in the last 540 million years. However, a recent study by Stanek *et al* (2006) says that GRBs are not possible in metal rich galaxies like our own. A proposal that a supernova or gamma ray burst had caused a mass extinction would also have to be backed up by

astronomical evidence of such an explosion at the right place and time.

It has been suggested that a supernova or gamma ray burst caused the End-Ordovician extinction.

11. *Continental Drift:* Movement of the continents into some configurations can cause or contribute to extinctions in several ways: by initiating or ending ice ages; by changing ocean and wind currents and thus altering climate; by opening seaways or land bridges which expose previously isolated species to competition for which they are poorly-adapted (for example the extinction of most American marsupials after the creation of a land bridge between North and South America). Occasionally continental drift creates a super-continent which includes the vast majority of Earth's land area, which in addition to the effects listed above is likely to reduce the total area of continental shelf (the most species-rich part of the ocean) and produce a vast, arid continental interior which may have extreme seasonal variations.

    It is widely thought that the creation of the super-continent Pangea contributed to the End-Permian mass extinction. Pangaea was almost fully formed at the transition from mid-Permian to late-Permian, and the "Marine genus diversity" diagram at the top of this article shows a level of extinction starting at that time which might have qualified for inclusion in the "Big Five" if it were not overshadowed by the "Great Dying" at the end of the Permian.

12. *Plate Tectonics:* This is the mechanism which drives many of the possible causes of mass extinctions, especially volcanism and continental drift. So it is implicated in many extinctions, but in each case it is necessary to specify which manifestations of plate tectonics were involved.

Many other hypotheses have been proposed, such as the spread of a new disease or simple out-competition following an especially successful biological innovation. But all have been rejected, usually for one of the following reasons: they require events or processes for which there is no evidence; they assume mechanisms which are contrary to the available evidence; they are based on other theories which have been rejected or superseded.

Postulated Extinction Cycles

It has been suggested by several sources that bio-diversity and/or extinction events may be influenced by cyclic processes. The best-known of these claims is the 26 to 30 million year viral cycle in extinctions proposed by Raup and Sepkoski (1986). More recently, Rohde and Muller (2005) have suggested that bio-diversity fluctuates primarily on 62 ± 3 million year cycles.

It is difficult to evaluate the validity of these claims except through reduction to statistical arguments regarding how plausible or implausible it is for the observed data to exhibit a particular pattern, as the causes of most extinction events are still too uncertain to attribute to them any specific cause let alone a recurring one. Much early work in this area also suffered from poor knowledge of the geological time scale (errors > 10 million years at times), though the time scale now available (uncertainties all < 4 million years) should be adequate for studying these processes.

While the statistics alone have been judged as sufficiently compelling to warrant publication, it is important to consider processes that might be responsible for a cyclic pattern of extinctions and future work may focus on trying to find evidence of such processes.

One theory, for which no real evidence exists, suggests that the extinction cycle could be caused by the orbit of a hypothetical companion star dubbed Nemesis that periodically disturbs the Oort cloud, sending storms of large asteroids and comets towards the Solar System. Another similar theory suggests that the Solar System's oscillations through the plane of the galaxy results in periods of comet showers. Other theories suggest geological instabilities that might allow heat to periodically build up deep in the Earth, which is then released through mantle plumes, periods of major volcanism and active plate tectonics.

If any of these theories are correct, then it is worth noting that both Raup and Sepkoski and Rohde & Muller predict another naturally caused mass extinction event within the next 10 million years.

Controversy

In 2005, Andrew Smith and Alistair McGowan of the Natural History Museum suggested that the apparent variations in marine bio-diversity may actually be caused by changes in the quantity of rock available for sampling from different time periods. The diversity of the marine life appears to be proportional to the amount of rock available for study. However, statistical analysis shows that only half of the apparent diversity modification can be attributed to this effect.

3

Management of Invasive Alien Species

Information Digest of Section

It discusses the recent cooperative work by international bodies to prevent the spread of invasive alien organisms and reduce their impacts. It also illustrates the variety of negative impacts that can occur when alien species become invasive in an area.

CBD Definitionsalien Species: a species, subspecies or lower taxon, introduced outside its natural past or present distribution; includes any part, gametes, seeds, eggs, or propagules of such species that might survive and subsequently reproduceinvasive alien species: an alien species whose introduction and/or spread threaten biological diversityIPPC definitions pest: any species, strain or biotype of plant, animal or pathogenic agent injurious to plants or plant productsquarantine pest: a pest of potential economic importance to the area endangered thereby and not yet present there, or present but not widely distributed and being officially controlled

Instruments Pertaining to the Management of Invasive Alien Species

The foremost international convention obliging its contracting parties to take action on invasive alien species is the Convention on Biological Diversity, which was adopted in 1992. The objectives of the convention are the conservation and sustainable use of biological diversity and the fair and equitable sharing of benefits arising out of the utilization of genetic resources. Article 8(h) of the CBD requires contracting parties to prevent the introduction of, control or eradicate those alien species which threaten ecosystems, habitats or species.

The International Plant Protection Convention, which came into force in 1952, also contains provisions applicable to invasive alien species when the species concerned are pests of plants or plant products. The IPPC aims to prevent the spread and introduction of pests of plants and plant products, and to promote appropriate measures for their control. The IPPC definition of a quarantine pest covers much, but not all, of what is considered as an invasive alien species under the CBD. Both definitions refer to any organism that is injurious to plants and that has an environmental impact. Although sometimes perceived as limited to the protection of agricultural and forestry plants, the IPPC covers all plants, including those found in natural and semi-natural habitats.

Many other international and regional instruments are relevant to managing the spread and negative effects of invasive alien species. Some 50 instruments or guidelines deal with particular aspects of the management of invasive alien species. Examples are listed below.

Major international instruments relating to invasive alien species:

- Convention on Biological Diversity
- International Plant Protection Convention
- Office International des Epizooties.

Instruments dealing with invasive alien species in the context of particular species groups include:

- Convention on the Conservation of Migratory Species of Wild Animals
- Agreement on the Conservation of African-Eurasian Migratory Waterbirds
- Convention on International Trade in Endangered Species of Wild Fauna and Flora.

Examples of instruments and programmes dealing with invasive alien species in the context of particular ecosystems are:

- United Nations Convention on the Law of the Sea
- United Nations Environment Programme Regional Seas Programme
- Ramsar Convention on Wetlands.

Instruments dealing with invasive alien species in the context of particular pathways for their introduction include:

- International Convention for the Control and Management of Ships' Ballast Water and Sediments
- FAO Code of Conduct for Responsible Fisheries.

Given the large number of such instruments, it is important for countries to be guided in how to develop national strategies on invasive alien species and understand how to put them into practice without running into conflict between one instrument and another. Measures to prevent the introduction of invasive alien species may, by their nature, be very trade restrictive. Therefore, members of the World Trade Organization must ensure also that any measures relating to invasive alien species are in accord with their obligations under international trade rules.

Recent Developments under the CBD Framework: In 2002, the CBD formalized a set of guiding principles on alien species that threaten ecosystems, habitats or species (*refer footnote on page* 7). These guiding principles comprise six general principles, three dealing with prevention, two with introduction of species and four with mitigation of impacts. The 15 subject areas are:

- precautionary approach
- three-stage hierarchical approach
- ecosystem approach
- the role of states
- research and monitoring
- education and public awareness
- border control and quarantine measures
- exchange of information
- cooperation, including capacity building
- intentional introduction

- unintentional introductions
- mitigation of impacts
- eradication
- containment
- control.

Many of the recommendations within these guiding principles correspond with provisions of the IPPC and its international standards for phytosanitary measures.

Other aspects of the CBD's programme of work on invasive alien species include a toolkit of prevention and management practices, a guide to designing legal and institutional frameworks, an analysis of gaps and inconsistencies in the international regulatory framework and an analysis of the ecological and socio-economic impact on island ecosystems and on inland water ecosystems. All of these relate to invasive alien species.

Recent Developments under the IPPC Framework: Since 1999, the governing body of the IPPC has devoted considerable effort to clarifying the role of the convention in relation to invasive alien species. In 2001, it formally determined that implementation of the IPPC is directly relevant to implementation of Article 8(h) of the CBD. It also agreed that many provisions and standards of the IPPC are directly relevant to, or overlap with, the (then interim) guiding principles of the CBD on invasive alien species. In particular, ISPM 11 [2001]: *Pest risk analysis for quarantine pests* was to be developed to address in detail the environmental risks of plant pests. The revised standard, ISPM 11 rev. 1: *Pest risk analysis for quarantine pests including analysis of environmental risks*, was adopted in 2003 and further revised and supplemented in 2004.

Also in 2003, ISPM 5: *Glossary of phytosanitary terms*, was supplemented with *Guidelines on the understanding of* potential economic importance *and related terms including reference to environmental considerations*. This clarifies that the IPPC can account for environmental concerns in economic terms using monetary or non-monetary values; market impacts are not the sole indicator of pest consequence. Thus the scope of the IPPC covers the protection not only of cultivated plants but also of uncultivated/unmanaged plants, wild flora, habitats and ecosystems.

Collaborative International Initiatives: Since 2001, the CBD and the IPPC have worked together in their efforts to help countries deal with the threat posed by invasive alien species. The aim is to avoid duplication of activities, contradictory approaches and a confusion of competences. A Memorandum of Cooperation has been established to formalize the arrangement between the two secretariats.

The CBD has recommended that its contracting parties consider ratifying the IPPC. It has invited collaboration in development of standards (such as those of the IPPC) when they could incorporate elements relating to threats to bio-diversity posed by alien species. There has been close collaboration between both conventions in the 2003 revision of ISPM 11. The CBD also works in partnership with the IPPC in information exchange endeavours such as the International Phytosanitary Portal and International Portal on Food Safety, Animal and Plant Health.

Range of Adverse Impacts: Cooperative international effort to foster the management of invasive alien species is particularly important because the problem is global. No single country can successfully control the spread and impacts of alien species when trade and travel promote the dispersal of such organisms and when their effects in a new territory may be unforeseen. The extent and range of adverse impacts makes cooperative action essential.

Although many alien species offer great benefits to a country (*e.g.* in agriculture, forestry, aquaculture), those species which become invasive can have devastating impacts. The negative impacts may be environmental through loss of bio-diversity, economic through a loss of production by affected species or the cost of control measures, health-related (*e.g.* when the invasive organism is a host or vector for disease) or political through effects on international trade, food security, water supply, regional stability, poverty, migration etc.

Invasive alien plants have a range of impacts on native biodiversity and ecosystems. They may compete with native taxa of flora, hybridize with genetically close species, alter the physical and chemical characteristics of soil, modify natural and semi-natural habitats, and propagate pests and diseases. The ecosystems most vulnerable to invasion are geographically and evolutionarily isolated ecosystems (islands, mountain ranges, lakes etc.) whose flora and fauna have evolved over millions of years.

Using the IPPC Framework: National phytosanitary authorities have a long experience in the assessment and management of biological risks related to the introduction of organisms. They possess the knowledge of how to deal with risks posed by plant pests. This existing infrastructure and know-how could be utilized by environmental authorities in their efforts to implement the guiding principles of the CBD.

The IPPC has been in effect for more than 50 years; over the past decade it has been developing a suite of international standards for the use of national plant protection organizations. IPPC-related activities are administered in many countries by agricultural authorities and CBD matters by environmental authorities. Communication between these authorities in relation to invasive alien species is important.

Section 1 suggests that, with its increased focus on environmental risks of plant pests, the IPPC provides a suitable framework to manage the introduction, control or eradication of invasive alien species that are plant pests. The infrastructure of national plant protection organizations is in place; the body of experience is well developed; relevant, user-friendly, international standards are in existence and more are being developed or revised; the necessary information exchange systems are established.

Gaps and Constraints in Mitigating the Effects of Invasive Alien Species: In addition to a lack of resources, common constraints at the national level include:

- lack of public, political and media awareness of environmental impacts of invasive alien species

- fragmented or outdated legislation that does not cover the full range of agricultural, environment, marine and public health concerns
- lack of a strategic approach and poor coordination between key departments and agencies.

The impacts of invasive alien species may extend to a group of neighbouring countries. Constraints on effective action at the regional level include lack of awareness, and poor coordination and cooperation between sectors and between neighbouring countries. Another problem arises with changes to regional free trade arrangements without consideration of the risks associated with liberalizing border controls and of management needs for invasive alien species already present within the free trade area.

Many pathways and vectors are still not adequately covered by international rules, guidance or codes of best practice. For example, seeds, food and other commodities moving in the course of development assistance, humanitarian or military operations fall outside the regulatory framework for conventional trade pathways. In 2004, the Conference of the Parties to the CBD identified a number of cases in which invasive alien species are not covered by the international regulatory framework.

The approach to invasive alien species is usually a defensive one. Existing frameworks do little to support "prevention through avoidance". There are few deterrents to the continued use of environmentally harmful species in some sectors and few incentives to promote the use of native species as an alternative to introduced species. There are few preventive rules to restrict exports of high-risk invasive plants or animals to countries where they are likely to be problematic. There are virtually no international early warning systems for invasive alien species as they affect wild species and natural or semi-natural systems.

Plants and animals that are invasive in their own right are covered under IPPC or OIE only if they qualify as plant pests or animal diseases. There is a lack of international standards for "environmental pests". Recent supplements to ISPMs do not extend the IPPC definition of "plant pest". The extension of some ISPMs to address environmental risks and costs will open up a new area for many plant health regulators.

More advanced tools are needed to deal with risk assessment and environmental assessment. Outside mainstream agriculture and forestry sectors information is lacking and impacts, except for particularly invasive species, are largely unknown. Existing tools do not usually cover ways to deal with species already present in a country and to determine factors of vulnerability for receiving environments.

Constraints on funding and gaps in institutional coordination are also areas that need addressing.

Recommended Steps: Section 1 suggests that common goals for all levels (international, regional, national) should include making best use of existing regulatory frameworks, strengthening cooperation between key organizations and targeting existing tools and resources more effectively to encompass bio-diversity-related impacts. Recommendations for action at the national level include development of policy and regulatory frameworks and the engagement of relevant community groups and entities:

- Policy and legal frameworks need to support broad objectives of protection: human, plant and animal health and life; species, subspecies and races against contamination, hybridization or extinction; native bio-diversity against impacts resulting from invasive alien species; and against biosecurity threats generally.
- Each country needs the legal basis, capacity and resources to support prevention, early detection and rapid response and eradication. Efficient use should be made of existing tools.
- Interagency cooperation and improved information-sharing between environment, agricultural and plant health sectors will facilitate the use of pest risk analysis for environmental risks, taking account of issues such as ecosystem dynamics, secondary pests and indirect impacts.
- The multiple roles of border and quarantine services in trade facilitation, food security, human health and

environmental protection should be recognized and capacity strengthened where possible.

- Industry, retailers, users and other groups should be actively engaged to ensure that measures selected are seen as legitimate and proportionate to the level of risk. The development of codes of conduct and of best practice should be promoted.
- Appropriate education and communication strategies may be tailored to different target audiences and groups, including enforcement personnel.
- Positive strategies need to be developed to encourage the use of native plant species and seeds of known local provenance in landscaping, countryside management, revegetation, erosion control, protected area management and international assistance programmes.

Convention and Invasive Alien Species

The Convention on Biological Diversity (CBD), which was adopted in 1992, incorporates provisions regarding those alien species which threaten ecosystems, habitats or species. The International Plant Protection Convention (IPPC), which has existed since the 1950s, aims to prevent the introduction and spread of plant pests. National plant protection services and the governing body of the IPPC, the Interim Commission on Phytosanitary Measures (ICPM), recognized that the aim of the CBD to prevent the introduction of alien species corresponds in large measure to the aim of the IPPC. Since 1999, the ICPM has been actively engaged in clarifying its role in regard to invasive alien species that are plant pests. In 2001, it determined that such species should be considered quarantine pests and should be subjected to measures according to IPPC provisions.

The ICPM also decided that IPPC standards should be reviewed to ensure that they adequately address environmental risks of plant pests. In 2003, the ICPM adopted supplements to two of the international standards for phytosanitary measures (namely *Glossary of phytosanitary terms* and *Pest risk analysis for quarantine pests*). These supplements elaborated on environmental considerations. To avoid conflicting developments within the IPPC and the CBD regarding invasive alien species and plant pests, the secretariats of the two conventions have established a Memorandum of Cooperation and developed a joint work plan as was called for by the Conference of Parties to the CBD at its seventh meeting.

In 2002, the Conference of Parties to the CBD formalized a set of guiding principles on alien species that threaten ecosystems, habitats or species. A comparison between these guiding principles and provisions of the IPPC shows strong correspondence and considerable overlap. This paper discusses the need for closer cooperation between environmental and phytosanitary authorities on a national and international level as well as future activities in regard to invasive alien species.

The United Nations Conference on Environment and Development (UNCED), commonly known as the "Earth Summit", was held in Rio de Janeiro in June 1992. One of the major results of the summit was the signature of the Convention on Biological Diversity (CBD). The three main objectives of the CBD (namely the conservation of biological diversity, the sustainable use of its components and the fair and equitable sharing of the benefits from the use of genetic resources) aim to maintain the world's environmental conditions in the face of its economic development.

Of the many detailed provisions in the CBD, one was of particular relevance to the international plant health community. Article 8(h) requires contracting parties to prevent the introduction of, control or eradicate those alien species which threaten ecosystems, habitats or species. During subsequent years, under guidance from its Subsidiary Body on Scientific, Technical and Technological Advice (SBSTTA), the CBD provided further advice on Article *impacts of alien species that threaten ecosystems, habitats or species* (CBD, 2002). These guiding principles were designed to assist all governments and organizations in developing effective strategies to minimize the spread and impact of invasive alien species.

The activities of the CBD in relation to alien species correspond to a certain degree with those carried out by the International Plant Protection Convention (IPPC) in relation to pests of plants and plant products. The IPPC is a multilateral treaty that was deposited with the Director-General of the Food and Agriculture Organization of the United Nations (FAO) in 1951, came into force in 1952, and was amended in 1979 and 1997 (FAO, 1997). Although the latest revision is not yet in force, its acceptance is imminent and countries have agreed to start implementation. The IPPC is designed to promote international cooperation in controlling pests of plants and plant products and in preventing their international spread, and especially their introduction into endangered areas.

Both the CBD and the IPPC are instruments established to protect certain conditions. On the one hand, the CBD aims to conserve biological diversity and, in the specific case of invasive alien species, to protect ecosystems, habitats or species. The IPPC, on the other hand, seeks to protect plants and plant products. The scope of the IPPC is not limited to the protection of agricultural plants, but covers all plants. Given that plants constitute a considerable and intricate part of our biological environment, it must be concluded that the IPPC's aim conforms, although on a smaller scale, to that of the CBD in relation to invasive alien species. Both conventions aim to address relatively similar problems.

International Plant Protection Convention, 1997 revision, Article I

"With the purpose of securing common and effective action to prevent the spread and introduction of pests of plants and plant products, and to promote appropriate measures for their control."

The remainder of this paper discusses the relationship between invasive alien species and quarantine pests as well as the activities of the IPPC in relation to invasive alien species. The relationship between the CBD's guiding principles on invasive alien species and the IPPC and its international standards for phytosanitary measures (ISPMs) is highlighted. The paper concludes by discussing how the CBD, the IPPC and national governments may proceed in furthering the implementation of both conventions.

Relationship between Invasive Alien Species and Quarantine Pests

The parameters for analysing the relationship between invasive alien species and quarantine pests are the agreed or official definitions of both conventions. The CBD defines an alien species as "a species, subspecies or lower taxon, introduced outside its natural past or present distribution; includes any part, gametes, seeds, eggs, or propagules of such species that might survive and subsequently reproduce" and further prescribes that an invasive alien species is "an alien species whose introduction and/or spread threaten biological diversity" (annex footnote 57, CBD, 2002). The IPPC, on the other hand, defines a [plant] pest as "any species, strain or biotype of plant, animal or pathogenic agent injurious to plants or plant products" and a quarantine pest as "a pest of potential economic importance to the area endangered thereby and not yet present there, or present but not widely distributed and being officially controlled" (FAO, 1997).

The IPPC definition of a quarantine pest covers much, but not all, of what is considered as an invasive alien species under the CBD. Both definitions cover any organism that is injurious to plants and that has an environmental impact (threatens biological diversity). Both definitions prescribe in different words that the environmental impact results from the organism's introduction and/or spread. It can be argued that most quarantine pests are invasive alien species and that those invasive alien species which are directly or indirectly injurious to plants are quarantine pests.

One marked difference between the definitions is that quarantine pests do not necessarily threaten biological diversity. For example, they may affect only agricultural or horticultural plants that are alien species in their own right.

Overlapping Mandates of International and Regional Organizations

The relationship between the CBD and the IPPC in relation to invasive alien species not only results in a partial overlap of the mandates of both conventions, but also draws several other international organizations into the picture. First and most important, the Agreement on the Application of Sanitary and Phytosanitary Measures (SPS Agreement) of the World Trade Organization (WTO) plays a prominent role. The SPS Agreement lays down trade-related rules concerning sanitary and phytosanitary measures. The basic aim of the SPS Agreement is to prevent countries from establishing arbitrary and unjust trade barriers by having unjustified import restrictions. Article 3 of the SPS Agreement lays down that "Sanitary or phytosanitary measures which conform to international standards, guidelines or recommendations shall be deemed to be necessary to protect human, animal or plant life or health, and presumed to be consistent with the relevant provisions of this Agreement..." (WTO, 1994). In order to establish which measures are necessary, the SPS Agreement relies on the work of three international standard-setting organizations in the fields of food safety, animal health and plant health. The IPPC has been designated in the SPS Agreement as the relevant international organization for the plant health field. This leads to the situation where trade-related rules for those plant pests which are also invasive alien species have to comply also with the SPS Agreement. In such instances, countries that establish phytosanitary import requirements would have to comply not only with IPPC and CBD provisions but also with the basic rules of the SPS Agreement. This overlap would also mean that the three international organizations need to cooperate to avoid any discrepancy between the provisions of the three agreements.

Regional plant protection organizations (RPPOs) may also be drawn into the international framework dealing with invasive alien species that are plant pests. According to Article IX of the IPPC, RPPOs function as the coordinating bodies in the geographical areas covered and participate in various activities to achieve the objectives of the IPPC. Consequently, RPPOs may be active in relation to invasive alien species. For example, the European and Mediterranean Plant Protection Organization (EPPO) has initiated an extensive work programme in this area.

Relation to Invasive Alien Species

The governing body of the IPPC, the Interim Commission on Phytosanitary Measures (ICPM), addressed the topic of biosafety at its second meeting (ICPM 2) in 1999. It concluded that the concept of invasive alien species had considerable implications for the IPPC and that coordination between government authorities at a national and international level was necessary to avoid conflicting views in different fora (FAO, 1999). As a result of these discussions, ICPM 2 created an informal open-ended working group to consider issues regarding GMOs, biosafety and invasive alien species. This group met in 2000 and, based on its recommendations, ICPM 3 decided in 2001 that:

- Species that may be invasive and that directly or indirectly affect plants or plant products should be assessed, monitored and managed, if necessary, according to IPPC provisions.
- Invasive alien species that are plant pests and that are absent from an area (or if present, are of limited distribution and under official control) should be considered quarantine pests and should be subjected to measures according to IPPC provisions.
- Implementation of the IPPC is directly relevant to implementation of Article 8(h) of the CBD.
- Many provisions and standards of the IPPC are directly relevant to, or overlap with, the (then interim) guiding principles of the CBD.
- Standards should be reviewed to ensure that they adequately address environmental risks of plant pests.
- A supplement to ISPM 11: *Pest risk analysis for quarantine pests* should be developed to address in detail the environmental risks of plant pests (FAO 2001).

ICPM 3 also decided that the IPPC Secretariat should cooperate with the CBD on matters related to invasive alien species. The IPPC Secretariat was asked to seek observer status for the IPPC with the CBD, to attend relevant meetings of the CBD and invite the CBD to attend relevant IPPC meetings. To avoid conflicting developments in different international organizations, ICPM 3 charged the IPPC Secretariat to cooperate with other standard-setting bodies to ensure that common areas of interest would be adequately covered.

The decisions of ICPM 3 on invasive alien species have continued to occupy the activities of the organization. In 2003, ICPM 5 adopted two ISPMs with direct relevance to invasive alien species and the protection of the environment:

- a supplement to ISPM 5: *Glossary of phytosanitary terms*. Supplement no. 2: *Guidelines on the understanding of* potential economic importance *and related terms including reference to environmental considerations* (FAO, 2003)
- a supplement to ISPM 11 [2001]: *Pest risk analysis for quarantine pests*. Supplement: *Analysis of environmental risks* (FAO, 2003).

At times the IPPC has been misinterpreted as referring to the protection only of cultivated plants. This misinterpretation may have arisen partly because the definition of a quarantine pest provides that only those pests which potentially cause economic damage may qualify as quarantine pests. The adopted guideline on the understanding of the term "potential economic importance" clarifies that the IPPC can account for environmental concerns in economic terms using monetary or non-monetary values and that market impacts are not the sole indicator of pest consequences. The clarification offered by the 2003 supplement to ISPM 5 assures that the scope of the IPPC covers the protection of cultivated plants in agriculture (including horticulture or forestry), uncultivated and unmanaged plants, wild flora, habitats and ecosystems.

The supplement to ISPM 11 is a technical standard intended to provide details regarding the analysis of risks of plant pests to the environment and biological diversity, including those risks affecting uncultivated/unmanaged plants, wild flora, habitats and ecosystems contained in the pest risk analysis (PRA) area.

In addition to these amendments to standards, the cooperation between the IPPC Secretariat and the CBD has developed. A Memorandum of Cooperation between FAO and the Secretariat of the CBD on cooperation between the secretariats of the IPPC and the CBD aims to promote synergy, to avoid overlaps and unnecessary duplication as well as to ensure effective cooperation in joint activities. A joint work plan has been developed as was called for by the seventh meeting of the Conference of Parties to the CBD (CBD, 2004).

The CBD's Guiding Principles on Invasive Alien Species and the IPPC

Comparing the aim and scope of the CBD and the IPPC reveals numerous overlapping elements. ICPM 3 noted that many provisions and standards of the IPPC are directly relevant to, or overlap with, the provisions of the (then interim) guiding principles of the CBD (Appendix XIII, FAO, 2001):

"Relevant areas include:

- providing legal and regulatory frameworks;
- building capacity and technical assistance for developing countries;
- assessing and managing potential plant pest risks;
- protecting areas that may be threatened by plant pests;
- applying measures to prevent unintentional introduction of plant pests;
- certifying that risk management procedures have been applied;
- assessing and managing the intentional introduction of organisms that may be pests of plants including claimed beneficial and biological control organisms;
- exchanging of scientific and regulatory information relevant to plant pests;
- cooperating between countries to minimize the impact of plant pests; and
- detecting, controlling, and eradicating pests in agricultural and wild flora."

In 2002, the sixth Conference of the Parties (COP) to the CBD adopted the guiding principles as non-binding recommendations intended to provide all governments and organizations with guidance for developing effective strategies to minimize the spread and impact of invasive alien species (CBD, 2002; *refer also footnote on page 7*). A detailed comparison between the recommendations made in the 15 guiding principles on invasive alien species and the IPPC provisions and its ISPMs may further clarify the relationship between invasive alien species and plant pests and the work of the IPPC and CBD.

Precautionary Approach

The precautionary approach in principle 1 refers to the Rio Declaration and the preamble of the CBD, which lays down that "... where there is a threat of significant reduction or loss of biological diversity, lack of full scientific certainty should not be used as a reason for postponing measures to avoid or minimize such a threat".

The IPPC does not have a similar provision. Indeed, its Article VII.2a specifies that phytosanitary measures shall not be taken without technical justification, which is understood to be a pest risk analysis. This apparent contradiction between the two conventions, however, becomes less evident on closer inspection. The use of the terms "significant reduction" and "lack of full scientific certainty" in the CBD's precautionary approach seems to imply that certain knowledge should be available. On the other hand, it is also understood by the IPPC that the availability of full scientific evidence is not always possible. ISPM 11 makes provision for uncertainties in the PRA process.

Three-stage Hierarchical Approach

A three-stage hierarchical approach to minimize the risk and spread of invasive alien species gives preference to the prevention of their introduction. In the event of an introduction, early detection and rapid action (*e.g.* eradication) are recommended. If this fails, the third stage is containment and control.

The approach can be considered as the underlying principle in phytosanitary policy. For decades, governments have operated their phytosanitary policy on the principle of preventing the introduction of quarantine pests, and if this fails, eradicating or containing them. This is also reflected in many provisions of the IPPC, although a detailed hierarchical approach is not explicitly specified.

Ecosystem Approach

Principle 3 recommends that measures to deal with invasive alien species should, as appropriate, be based on the ecosystem approach. This approach may be described as a strategy to protect complex and dynamic plant, animal and microorganism communities and their non-living environment, which together interact as functional units, through integrated management of land, water and living resources (CBD, 2000).

Relationship between the CBD's Guiding Principles and the IPPC and its Standards

Guiding principle, CBD	Corresponding IPPC provisions and/or ISPMs
1 Precautionary approach	Article VII.2a (potential contradiction)
	Relevant ISPMs: ISPM 1 (partial application)
2 Three-stage hierarchical approach	General aim of IPPC (full application)
3 Ecosystem approach	Relevant ISPMs: ISPM 11 supplement on environmental risks; ISPM 5 supplement on potential economic importance
4 The role of states	General aim of IPPC (full application), Article IV
	Relevant ISPMs: ISPM 3, 6, 17, 19
5 Research and monitoring	Article IV.2b and Article IV.3b
	Relevant ISPMs: ISPM 6, 8
6 Education and public awareness	not covered (Article IV.3a provides very limited coverage)
7 Border control and quarantine measures	especially Article V and Article VII
	Relevant ISPMs: ISPM 1, 7, 12, 13, 14
8 Exchange of information	especially Article VIII
	Relevant ISPMs: ISPM 1, 6, 8, 13, 17, 19
9 Cooperation, including capacity-building	especially Articles VIII, IX and XX
10 Intentional introduction	especially Article VII
	Relevant ISPMs: ISPM 1, 2, 3, 11
11 Unintentional introductions	especially Article VII
	Relevant ISPMs: ISPM 1, 2, 11, 14
12 Mitigation of impacts	Article I
13 Eradication	Article I
	Relevant ISPMs: ISPM 9
14 Containment	Article I
	Relevant ISPMs: ISPM 4, 9, 10
15 Control	Article I
	Relevant ISPMs: ISPM 14

Definitions of CBD, IPPC and ISPM in Accompanying Text

The fact that the IPPC is a convention deposited with an agricultural organization (FAO) and that countries have mostly regulated pests of agricultural and horticultural crops has led to the perception in the environmental community that the IPPC does not address environmental concerns. As described earlier, in 2003, the ICPM developed and adopted two supplement ISPMs to give guidance to countries on how to address environmental concerns. Of these, the supplement to ISPM 11 deals explicitly with the analysis of environmental risks and incorporates the ecosystem approach into the standard on pest risk analysis for quarantine pests.

The Role of States

This guiding principle calls for states to recognize that activities within their jurisdiction or under their control, such as intentional and unintentional introductions, may pose risks to other states. Guiding principle 4 stipulates that states should take actions to minimize the spread and impact of invasive alien species. This would include the identification of invasive alien species or species that could become invasive as well as providing information on such species to other states.

The IPPC recognizes the role of states in combating plant pests. In fact, the IPPC goes further by requesting the establishment of adequate structures in states. Especially Article IV of the IPPC lays down general provisions for the organizational arrangements for national plant protection, such as the establishment of an official national plant protection organization (NPPO) and its functions in each state. Several ISPMs give guidance on the role of states in their actions to minimize risks. ISPM 3: *Code of conduct for the import and release of exotic biological control agents* is important in relation to the intentional introduction of organisms, and ISPM 6: *Guidelines for surveillance* and ISPM 19: *Guidelines on lists of regulated pests* to the identification, surveillance and listing of pests, while ISPM 17: *Pest reporting* describes responsibilities of states in reporting the occurrence, outbreak and spread of pests.

Research and Monitoring

This guiding principle states that it is important for states to conduct research on and monitoring of invasive alien species in order to increase the knowledge about such species and their status in the country.

The IPPC also specifically addresses research and monitoring. Article IV.2b of the IPPC attributes the surveillance of cultivated plants and wild flora to the responsibilities of NPPOs. Article IV.3b prescribes that each state shall make provisions concerning research and investigations in plant protection. Practical guidance in relation to monitoring is provided in ISPM 6: *Guidelines for surveillance* and ISPM 8: *Determination of pest status in an area.*

Education and Public Awareness

Guiding principle 6 attributes importance to public awareness in the management of invasive alien species. It recommends that states should promote education and public awareness of the causes of invasion and the risks associated with the introduction of alien species. In cases of mitigation measures, such as control or containment programmes, this should be done in a way to involve local communities and appropriate interest groups.

The IPPC does not have similar provisions. It partly covers the provisions in Article IV.3a, which prescribes that states should make provisions for the distribution of information within their territory. Public awareness initiatives, however, are carried out by many NPPOs as a regular and useful tool in preventing the introduction of pests or in involving the public in specific surveillance programmes.

Border Control and Quarantine Measures

This guiding principle recommends that states should implement border controls and quarantine measures to minimize the risks of introduction of alien species that are or could become invasive. The quarantine measures should be based on risk assessment, and existing appropriate government bodies should be strengthened as necessary to implement the measures.

Provisions concerning quarantine measures and border control are the main substance of the IPPC. In particular, Articles V and VII deal with phytosanitary certification and requirements in relation to imports, respectively. Several ISPMs provide practical guidance in relation to border controls and quarantine measures. The most important are: ISPM 1: *Principles of plant quarantine as related to international trade*, ISPM 7: *Export certification system*, ISPM 12: *Guidelines for phytosanitary certificates*, ISPM 13: *Guidelines for the notification of non-compliance and emergency action* and ISPM 14: *The use of integrated measures in a systems approach for pest risk management*.

Exchange of Information

Provisions regarding information exchange on alien species are laid down in principle 8. It recommends the development of information systems in regard to relevant biological information on alien species as well as the dissemination of information. Information on import requirements for alien species should be made available to other states.

Perhaps one of the most important differences between principle 8 and the IPPC provisions is that many of the IPPC provisions on information exchange are obligatory to contracting parties. Article VIII specifies that contracting parties shall cooperate in the exchange of information on plant pests and that they shall designate an official contact point through which the information exchange is facilitated. Several articles of the IPPC specify which type of information contructing parties are obliged to exchange and with whom. The IPPC Secretariat has put into operation the International Phytosanitary Portal, an Internet-based information system and network, which facilitates international information exchange. Several ISPMs also provide guidance on information exchange, especially ISPM 17: *Pest reporting*, which describes detailed reporting responsibilities. Other ISPMs dealing partially with information exchange are ISPM 1: *Principles of plant quarantine as related to international trade*, ISPM 6: *Guidelines for surveillance*, ISPM 8: *Determination of pest status in an area*, ISPM 13: *Guidelines for the notification of non-compliance and emergency action* and ISPM 19: *Guidelines on lists of regulated pests*.

Cooperation, Including Capacity Building

This guiding principle points out that a state's response to minimizing the spread and impact of invasive alien species not only may be applied internally within the country but also may require a bilateral or multilateral approach with other countries. Cooperative efforts may include the development of programmes to share information and the establishment of bilateral or multilateral agreements to regulate trade in certain alien species, as well as cooperation in research and its funding. Capacity-building programmes for states that lack expertise and resources are advocated. Such programmes may involve technology transfer and the development of training programmes.

The IPPC also calls for cooperation and capacity building to achieve its aims. Multilateral cooperation is especially advocated in Articles VIII and IX of the IPPC. Capacity building has an especially high profile in the IPPC. In Article XX, contracting parties agree to promote technical assistance to facilitate implementation of the convention. This is reflected in the work of the IPPC Secretariat which devotes considerable resources to technical assistance. The development and practical application of the Phytosanitary Capacity Evaluation (PCE) module, a tool to evaluate the phytosanitary capacity of countries, has increased considerably the understanding of technical needs of developing countries.

Intentional Introduction

Guiding principle 10 provides recommendations regarding the intentional introduction of alien species into countries or into new ecological areas within a country. The principle recommends that such intentional introductions should take place only after they have been evaluated and authorized. A risk assessment should be part of the evaluation and the authorization should be based on the precautionary principle. Furthermore, principle 10 recommends that the burden of proof that a proposed introduction is unlikely to threaten biological diversity should be with the proponent of the introduction or be assigned as appropriate by the recipient state.

The IPPC covers requirements in relation to imports especially in Article VII. Specifically, Article VII.5 addresses the intentional import of pests and regulated articles for research, education or specific use. In cases of intentional import, special and adequate safeguards should be established to prevent the "escape" of the pest.

Particularly relevant to the intentional introduction of alien species, ISPM 3: *Code of conduct for the import and release of biological control agents* describes the responsibilities of authorities of governments, importers and exporters in relation to the importation of exotic biological control agents capable of multiplication. Other IPPC standards relevant to this guiding principle are ISPM 1: *Principles of plant quarantine as related to international trade*, ISPM 2: *Pest risk analysis* and ISPM 11 [2004]: *Pest risk analysis for quarantine pests, including analysis of environmental risks and living modified organisms*.

Unintentional Introduction

This guiding principle recommends that every state should have in place provisions to prevent unintentional introductions of invasive alien species. Such provisions could incorporate legislative measures and the establishment or strengthening of institutions. Guiding principle 11 also mentions the most common pathways for unintentional introductions, such as agriculture, forestry, shipping or tourism. It recommends that environmental impact assessments of such activities should consider unintentional introductions of invasive alien species and that risk assessments should be carried out for these pathways, where appropriate.

Unintentional introductions are probably the most common problem in the phytosanitary field. The IPPC addresses in several of its articles the provisions recommended in the guiding principle 11. Most relevant is Article VII with its requirements in relation to imports. ISPMs of relevance to guiding principle 11 are ISPM 1: *Principles of plant quarantine as related to international trade*, ISPM 2: *Pest risk analysis*, ISPM 11 [2004]: *Pest risk analysis for quarantine pests, including analysis of environmental risks and living modified organisms* and

ISPM 14: *The use of integrated measures in a systems approach for pest risk management.*

Mitigation of Impacts

Guiding principle 12 (as well as the three following guiding principles) deals with the mitigation of impacts once the establishment of an invasive alien species has been detected. Mitigation measures, which should be initiated at the earliest possible date, may be eradication, containment or control programmes that are safe to humans, the environment and agriculture, as well as ethically acceptable to stakeholders.

Principle 12 also recommends that, consistent with national policy or legislation, an individual or entity responsible for the introduction of invasive alien species should bear the costs of control measures and biological diversity restoration where their failure to comply with the national laws and regulations is established.

The IPPC does not go into detail of what should happen if a quarantine pest has been introduced into a country. Article I specifies that one of the purposes of the IPPC is to promote appropriate measures to control pests. It is, however, common practice for countries to implement mitigation measures once a pest has been detected. In relation to the safety of the mitigation measures and their ethical acceptability to stakeholders, the IPPC has no provisions equivalent to those of the CBD's guiding principles on invasive alien species. Also the "polluter pays" principle as laid down in principle 12 is not reflected in the IPPC or any of its standards.

Eradication

This guiding principle recommends dealing with the introduction and establishment of invasive alien species by eradication, where feasible. Eradication is best carried out in the early stages of an invasion, and community support is often essential for the success of an eradication campaign.

The eradication of pests is implied in Article I of the IPPC. More detailed provisions concerning the establishment of pest eradication programmes are provided in ISPM 9: *Guidelines for pest eradication programmes*, which specifies administrative and technical components of such programmes.

Containment

Guiding principle 14 suggests that containment, or limiting the spread, of alien invasive species may be an appropriate strategy where eradication is not feasible.

As in the case of eradication, the containment of pests is implied in Article I of the IPPC. Several ISPMs provide partial guidance on limiting the spread of pests. These are ISPM 4: *Requirements for the establishment of pest free areas*, ISPM 9: *Guidelines for pest eradication programmes* and ISPM 10: *Requirements for the establishment of pest free places of production and pest free production sites.*

Control

In cases where eradication and/or containment has failed, control measures are the last step in efforts to minimize the impact of alien invasive species. Guiding principle 15 recommends that control measures focus on reducing the damage caused by invasive alien species as well as reducing their number. This principle highlights the use of integrated management measures.

Article I of the IPPC defines that promoting appropriate measures for the control of pests is one of the purposes of the convention. Practical guidance on the control of pests through integrated measures is given in ISPM 14: *The use of integrated measures in a systems approach for pest risk management*, although this standard is focused on the development of pest risk management options for the import of plants and plant products.

Discussion

As presented above, there are considerable overlaps in the work of the CBD on invasive alien species and the IPPC activities in regard to plant pests. To avoid inconsistencies and duplication of effort and to achieve effective action, we should consider the following questions:

- How can plant health authorities assist environmental authorities in realizing their objectives?
- How should national and international environmental and phytosanitary bodies cooperate in future in dealing with invasive alien species?
- What are important future activities in the field of invasive alien species?

These issues are discussed below:

The role of phytosanitary authorities in the work on invasive alien species

For almost 100 years, phytosanitary authorities worldwide have carried out the important task of preventing the introduction of quarantine pests. An efficient infrastructure (such as border controls, national surveillance programmes, technical and scientific institutions, as well as export-oriented certification programmes) has been established to achieve the tasks of phytosanitary authorities. The long experience of phytosanitary authorities in the assessment and management of biological risks related to the introduction of organisms provides these authorities with the knowledge of how to deal with risks posed by plant pests and invasive alien species that are plant pests.

On a national level, this existing infrastructure and know-how should be utilized by environmental authorities in their efforts to implement the guiding principles of the CBD. Such utilization would have considerable advantages for governments since existing structures and know-how would be used without significant new investments and a duplication of activities would be prevented.

Cooperation between International Bodies

The argument for close national cooperation between environmental and phytosanitary authorities also holds true on an international level. To avoid duplication of activities, contradictory approaches and a confusion of competences, the IPPC and the CBD should work closely together in relation to invasive alien species. Such a close cooperation need not be limited to the secretarial levels of both conventions, but may also include joint activities of the relevant governing bodies. This could be achieved either through a declaration by the CBD that the IPPC is a competent authority for the development of technical standards on invasive alien species that are pests of plants, including invasive plants, or through the establishment of a formal inter-organizational working group developing recommendations for invasive alien species.

Improving the cooperation between international organizations should not be limited to the IPPC and the CBD. The SPS Agreement of the WTO should also be included in such a closer collaboration. Measures to prevent the introduction of invasive alien species may, by their nature, be very trade restrictive. A close cooperation between the CBD, the IPPC, the SPS Agreement and possibly other international organizations could certainly help to achieve the objectives of these instruments without restricting trade unnecessarily.

Closer cooperation between international organizations depends on the efforts of national governments because it is ultimately their responsibility to determine the policy of the international organizations to which they belong. A number of countries have not ratified or accepted the CBD and/or the IPPC. If countries wish to influence the policy of these organizations their primary objective should be to ratify them.

CBD matters are administered in many countries by environmental authorities and IPPC-related activities by agricultural authorities. Hence, communication between these authorities in relation to invasive alien species is important. If governments are to address matters related to invasive alien species and plant pests in the CBD and the IPPC in a consistent way, it would be advisable for national coordination strategies to be developed.

Future Activities

It is of importance to states that different international agreements partly covering the same substance do not provide contradictory approaches or room for conflicting interpretations. Situations where national governments by implementing the provision of one international agreement may violate the provisions of another should be avoided. Thus it would be advisable if joint activities could be conducted to analyse the guiding principles of the CBD and the IPPC and its standards in order to resolve possible conflicting approaches and to agree on a common understanding and terminology.

The success of the CBD and the IPPC in protecting plants and the environment is very much dependent on the implementation of their provisions by national governments. The prevention of the spread of invasive alien species and plant pests is primarily an international approach in which countries must cooperate to prevent the natural or man-facilitated spread of such organisms. For many developing countries and especially for least-developed countries, however, the protection of the environment may not be located high on their list of national priorities. These countries may need to use their scarce resources to establish basic economic conditions taken for granted in the developed world. Thus the provision of technical assistance for developing countries should be seen as one of the priorities for the CBD and the IPPC to further the implementation of their provisions. A close cooperation between the CBD and the IPPC on technical assistance activities in relation to invasive alien species and plant pests would maximize the use of resources provided for this purpose and utilize FAO's experience in providing technical assistance.

4

Herbal Sources of Tibetan Medicines

Protecting the Himalayan varieties used in Tibetan medicines, exploring growing them as crops, and evaluating the healing properties of plants that grow elsewhere. Tibetan medicine is celebrated as a source of sustainable and affordable healing preparations that are effective without lasting negative side effects. Our goal is to give people all over the world access to this unique insight into human well being.

Along with the need to provide training for a new generation of physicians and herbal pharmacists, and to translate the Tibetan medical so that students can begin their training without first having to learn to read Classical Tibetan, the main obstacle to wider use of Tibetan medicine is the limited supply of many herbs used in compounding the medications, some of which are already endangered by excessive non-professional harvesting.

His Holiness The Dalai Lama has warned that the health of the billions of people in the world cannot be sustained by medicines made with rare and endangered Himalayan plants. Even now, when Tibetan medicine has hardly begun to be practiced in Western countries, the herbs needed for making many of the medicines are in short supply.

For example, the Men-Tse-Khang pharmacy in Dharamsala, India—the most respected source of Tibetan medicines—turns away all requests for medications except those accompanied by a prescription written by one of their own doctors. They will not accommodate even orders from their own graduates, those who have emigrated from India to other countries, explaining that to do so would deplete the supply of medicines needed in their branch clinics in India and Nepal.

Three different approaches to this scarcity of medicines are being explored. The most obvious and the most urgent is to preserve whenever possible the plants and animals currently used in Tibetan medicines, by protecting the ecologies that support those organisms, planting some species to increase the supply, and developing resources like seed banks to insure these species against extinction.

A second strategy is to extend the area where the herbs grow by planting in the wild and by finding ways of cultivating some species, as a way of preserving those that cannot be preserved in the wild, and as a way of increasing supplies.

The third is to evaluate the medicinal qualities of plants in other areas. Tibetan doctors have always used many imported materials in making their medicines, and believe that using substitute materials with similar therapeutic action is entirely appropriate when the ingredients of choice are unavailable. Tibetan physicians can identify plants indigenous to other areas that are suitable for use as substitutes for ingredients in traditional formulas and in developing new preparations.

Herbal medicine systems from other cultures will be useful in identifying promising species. Especially promising are two other systems which have ancient connections with Tibetan medicine and are still practiced: Indian Ayurvedic medicine and Chinese herbal medicine. Many of the herbs used in these systems are already used in Tibetan medicine.

The research needed for effort seems endless. The newsletter of the Chakpori Tibetan Medical Institute in Darjeeling, India, comments on some of the main issues: "To know which plants are vulnerable, endangered, or close to extinction, detailed studies have to be made. Also, the cause of rarity should be investigated (*e.g.*, environmental causes, over exploitation by non-expert gathering, etc.). An important decision has to be made to replant not only the economically valuable plants but also the plants that are environmentally valuable and supportive of the total environment.

"Replanting sites should have the same characteristics as the original habitat. In traditional Tibetan medical texts, the side of the mountain, altitude, climatic condition, soil composition, etc. are mentioned [as important influences on the properties and potency of the herbs].... research findings are essential guidelines for replanting the herbs in their natural environment. The size of the replanting area is also crucial because of genetic diversity and the problems of insular ecology. Monocultures, like the usual big plantations, should be avoided: mass plantations of certain species will not have the desired quality or power."

Here on this page we offer links to Web sites relevant to preserving the plant species used in making Tibetan medicines, and increasing the supply of those herbs or finding substitutes, along with books and audio tapes on Tibetan herbs and related topics. Another page, Medical Research, Tibetan Style, gives resources relevant to other aspects of the immense task of preserving Tibetan medicine and making it available to the world.

New Dimensions In Cultivating Medicinal Plants

Biodynamics

is the science of life forces, a recognition of the basic principles at work in nature. Bio means "life" and dynamics refers to "energy".

Hence Biodynamic agriculture suggests working with the energies which create and maintain life.

Humanity has created a situation where we can truly say that our Earth is ill. There is not a single place on this planet, apart from a few very remote areas which have not been contaminated with chemical pesticides, or even with fertilisers. This condition a serious affair and has come about only in the last fifty years.

Chemical supplied by the industry, and, used in agriculture, have been a result of War. Most of the fertilisers - especially the urea fertilisers-were developed during War time as explosives. Since that time, they have been used extensively in agriculture. In fact, the whole agriculture of this century has drawn upon natural reserves of humus and soil, which have been accumulated by Nature through many decades of natural vegetation growths.

Having produced those chemicals, and we know very little about their consequent effects on the environment. Chemical fertilisers are very soluble salts, which change the nature of the soil, waters, and substantially alter the ecology of the plants.

Using chemical fertilisers is similar to a non-nutritious human diet. Can plant nutrition be reduced to only a couple of minerals?

Fertiliser-based methods that reduce plant nutrition to a set of minerals imply that one can disconnect the health of a plant from its surrounding ecology, including insect pests. This approach is one that treats illnesses according to symptoms and not the cause. In this way it is similar to the modern allopathic method of treatment.

Many modern medicines are based on symptomatic treatment. The same mode of thinking has been applied in agriculture: there is a pest and the pest has to be killed, no matter what happens around it. It is not realised, that often the pest has its origin in the malnutrition of the soil. Improving the quality of the soil leads to stronger, healthier plants which in turn are more disease and pest resistant. This fact suggests that if plants were healthier, pesticides would not be necessary.

The earth is a living body. Whatever we do at this part of the planet has consequences at another. Even the people living in the Arctic have been tested positive for traces of chlorides and DDT in their bodies, although there is no farming in those remote regions of the planet. Why? Somebody in Africa is spraying DDT on his cotton crops. The rains wash the pesticidine into the river which in turn carries it into the ocean. Oceans are the circulation body of the earth; all their currents reach to every corner of the globe including the poles... and the circle continues.

We need to change the way of farming, not only of medicinal plants but also of our food, fruits and grains. Biodynamics recognises that soil itself can be alive. If we enrich the soil with life-sustaining properties the plants will carry those life forces. That, in turn, will give us life as well. Therefore, one of Biodynamics fundamental efforts is to build up stable humus in our soil through compositing.

In India, I have worked over the years with more than 1000 farmers who are now cultivating cotton in the biodynamic way. Along with cotton they cultivate a whole range of crops, such as wheat, maize, peas, soya beans, bananas and sugarcane.

In South India, we have introduced these agricultural principles in an area of about 7000 acres. Also tea is cultivated organically. Likewise, we have taken up this method in some of the Darjeeling tea gardens.

Origin of Biodynamic Agriculture

The biodynamic system was developed in the 1920s in Europe. It draws extensively on traditional agricultural knowledge and wisdom which was present in the traditional farming communities. Rudolf Steiner (1861-1925), an Austrian philosopher, scientist, artist, educator, and the founder of antroposophy, developed this agricultural system. Biodynamic cultivation has developed into a worldwide movement since 1922.

The products from this type of agriculture are identified in the market with the Demeter symbol. On seeing this symbol affixed to a packet of carrots the consumer in Europe, the United States, Australia or New Zealand, knows this product was cultivated according to this biodynamic practice, an agricultural system which aims at quality.

Informed consumers in the industrialised countries have now realised the importance of organic and biodynamic food, and are willing to pay a premium price for food that is 100% free from any pesticides.

Principles of Biodynamic Cultivation

Quality or Quantity

The main aim of biodynamic cultivation is to improve the quality of either food, fibre or medicinal plants. Of course, it also aims at improving the yields but, most importantly, it concentrates on quality.

In this century, more and more chemical fertilisers have been developed to produce maximum yields. The main objective has been to increase the yields. The reason is simple: bigger yields bring more money. Most vegetable producers have been cheating the consumer by selling them water for the price of vegetables. For example, dried carrots which have been chemically fertilised show less weight than organic carrots.

Adding fertilisers, which are salts, to plants, means that the plants draw more water from the soil in order to balance the ratio of fibre to salt inside. Consequently, as the plant cells expand and swell with water, the ratio of proteins and even vitamins decreases, and the essential oils, which have important medicinal properties, deteriorate. It has been scientifically proven that the amount of vitamins is less in chemically fertilised products. There is an imbalance of vitamins because the fertilised plants draw only one kind of nutrient and not all the trace elements. Hence, it does not produce the entire range of vitamins, and its whole value changes.

Quality of Medicinal Plants

What we want to achieve in biodynamic farming of medicinal plants, is the cultivation of the inherent quality of the plant. That is the very subtle active principle, which the plant has in very minute quantities. Instead of substance, here we are dealing with essential oils, which are almost not substantial, sometimes present only in parts per million.

What is quality, and how can we provide quality in food or in medicinal plants? What are the indicators for quality in medicinal plants? Quality comprises certain active principles, such as taste, juice, colour, seed content in the fruits, nutritional value, the vitamin and protein content. The question is, do we know how to evaluate subtle medicinal properties?

How do we define the taste, the aroma, the smell of flowers? How do we increase the scent of a rose, and how do we make oranges more sweet? How do we increase the content of active principles in medicinal plants? Does science know how to produce the best quality of food? I am not an agricultural scientist, but I question whether we know, for example, how to produce maize with the best content of protein or vitamins. The answer, I think, is no.

Modern agricultural sciences are based on how to work with yields. We know how to produce quantity but we don't know how to produce quality. We are facing the tragedy that roses don't have any smell any more; oranges have lost their taste; tomatoes are not sweet any more; carrots are bitter, and cultivated medicinal plants have less active principles.

Characteristics of the Himalayan Flora and Fauna

Each plant develops in a particular ecological zone. Here in the Darjeeling Hills we meet with this fantastic possibility of having land from very low elevations up to very high summits. Additionally the climate provides a lot of warmth and moisture. This combination of moisture and warmth creates very specific kinds of chemicals in the plants.

In high altitude mountains the principles are mainly dry and cold, with a lot of light. In contrast, the valleys are full of moisture, heat with a lot of vegetation and rather dark forests. Those extremes produce substances that are very different from each other.

In the higher regions of the Himalayas plants grow on very dry alpine soils. These, in many cases, are sandy and not very fertile. The cold climate limits the plants' growth. On lower elevations, the plants produces a lot of leaf, and in higher altitude they prominent with the flowers.

Polarities in Biodynamics

There is a polarity between yields and quality, between substance and un-substance, between big quantity of products and very small quantities of subtle ethereal oils. Regarding the definition of quality, a polarity can be observed between something very physical, such as the weight, substance - and something very subtle, such as ethereal oils.

In biodynamic farming we are trying to rediscover what is important in agriculture.

Earth and Sun

The entire plant life is a product of the earth and the sun- as some people say, the Mother Earth, and the Father Sun. In many cultures, the sun is masculine and the earth is feminine.

Sun	Earth
Warmth, Heat	Coolness, minerals,
Seasonal Rhythm	Gravity, Water,
Space	Soil
Light	**Darkness**

The Principle of Light

The colours of the beautiful high altitude flowers are a powerful manifestation of light. Colour is the resistance of substances to light, the tension between the light and the earth. The alpine regions are full of light.

This leads to very special structures of medicinal plants: the flowers are very colourful and bright, the leaves are very delicate - and with a lot of medicinal value. With hardly any substance alpine medicinal plants tend to be small but very powerful. The light that they accumulate is all transformed into essential oils. The more light, the more essential oils are produced. These medicinal plants pass through the whole winter with very little water. They are processing all this light, and they have to concentrate everything into essential oils.

The tannin in the tea leaf, for example, is nothing but concentrated sunlight. It is this sunlight which activates our senses, working on very subtle principles.

Using Polarity Principles in Biodynamic Agriculture

An example: Curcure (Equisetum arvense L.) is a plant without leaf. It consists only of stem. It grows in shady places, and is very rich in silica. The amount of silica, almost 90%, is one of the highest in the plant world. Silica is a mineral, belonging to the earth, but it lets the light go through. I call silica the forgotten nutrient.

Where do we find silica? In sand. Sixty or seventy per cent of the earth's crust is silica. And if you want to know how silica is connected to life, just look at a window: the glass is pure silica. There is silica in spectacles, silica in the eyes and skin, in the hair - and in leaves. Silica has an affinity to light. In fact, it draws light.

With silica as a carrier of light, we can make use of the polarity of light and darkness in agriculture. Equisetum arvense contains a lot of silica, and therefore belongs to the realm of light. On the contrary, fungus belongs to the realm of cold, moisture and darkness. We can use a product from one realm to control an imbalance in another. Equisetum arvense is an excellent fungicide.

Biodynamic Preparations

Out of his own insight, Rudolf Steiner introduced what is known as biodynamic preparations. Naturally occurring plant and animal materials are combined in specific recipes in certain seasons of the year and then placed in compost piles. These preparations bear concentrated forces within them and are used to organise the chaotic elements within the compost piles. When the process is complete, the resulting preparations are medicines for the Earth which draw new life forces from the cosmos. Effects of the preparations have been verified scientifically.

One of the preparations is made from silica powder, processed in a special way. It is applied to the plants at the time when they are producing the part that shall be harvested. In case of medicinal plants, if, for example, the root is harvested for medicinal purposes, silica is applied when the root is forming. The same applies for the flower, leaf, etc.

This silica powder preparation helps to increase the concentration of silica in the plant. With silica, the plant increases its rate of photosynthesis, draws more light, and produces more of those finer aspects. The medicinal value of plants cultivated under such conditions will be naturally higher and their pests lesser.

Silica is an excellent remedy for the foggy Darjeeling climate. The monsoon mists hold back the plant, whereas silica will activate the plant growth. Too much shade also creates a situation for fungus, which silica counteracts. In this way one can balance an unfavourable climatic situation with biodynamic preparations.

One of the biodynamic preparation is made from quartz crystals: the crystal form of silica. These crystals have geometric forms, perfect hexagonal shapes. The geometry shows another polarity of form and formlessness: in silica we have these beautiful shapes, and in fungus we have amorphous shapes, without form.

Geometric Principles and Medicinal Values

Alpine flowers are of the most perfect shapes, like stars: they have the most beautiful, artistic quality. The same applies for this quartz. These perfect shapes are based on the same geometric principle. It belongs to the heights, near the light, close to the stars. In pentagonal and hexagonal flowers we find the same geometrical principles as in crystals.

The differences between the root and the flower of such alpine medicinal plants reveal the polarity that we are talking about. The flower is very nicely shaped, though every species is different. But why do we not study the roots to differentiate those plants?

Because they are all more-or-less the same. There is no differentiation at the root level, and yet there is this high differentiation at the flower level.

How do these plants know that they have to grow like this? Why are they shaped and coloured like this? For example, this plant has five petals, five sepals, five stamens, so it is shaped by the principle of five, but where does it come from? It is these principles that form the plant, and along with it its medicinal values.

We can release this formative principle in our plants and understand how it starts shaping them and bringing forth these fine aspects of scent, aroma, taste, colour and medical properties.

In comparison, in tropical plants the principle of flower has been pushed into the leaf. Tropical climate produces fleshy plants that would never grow in the Alpine regions. These plants need moisture. It is warmth, which, combined with light, manifests in these plants in the leaves instead of in the flower.

Many plants, where this principle of warmth pushing into the leaf operates, develop a lot of poison. When this principle, that belongs to the flower, has been pushed as far as into the root, it colours even the roots. Colour at the root level means that what belongs to the flower has been pushed deep inside the plant.

Cosmic Integration

The Zodiac Principles

The ultimate fine-tuning of biodynamic principles lies in harnessing cosmic influences for cultivation. Only at particular times of the month or year are the cosmic influences are most supportive to the growth of a particular part of the plant.

The cosmic factor that determines a month is the moon. The movement of the moon in relation to the Zodiac is most interesting. The Zodiac symbols are Greek in origin. The Chinese developed a different system, and the Tibetans created their own amalgamation of the Indian and Chinese systems.

All the different systems have twelve constellations, though represented by different archetype figures and animals. Within these twelve signs there are four groups of three constellations each which have the same qualities. They are related to the four elements, earth, water, fire, air. These four elements can be placed in relation to the four parts of the plant: the root, the leaf, the flower and the fruit.

- The roots are associated with the earth: there is no base and ground without earth.
- The leaves relate to the water element: the more water, the more leaf.
- The flower corresponds to air and light: there is no light without air. (There is no light on the moon because there is no atmosphere.)
- Fruits are associated with fire: there is no ripening without warmth.

Cultivating the soil on specific days means harnessing the cosmic influences for the particular plant. Recent research has shown that in one month there are three periods which are variously beneficial for the root, the leaf, the flower and the fruit. These period are governed by the moon cycle. The different parts of the plant are stimulated, one at a time, three times in a month. Researchers discovered this by planting radishes every day and then observing their growth. Radishes from one day would be of very of good quality while those from another day would be smaller and less healthier. In between would be radishes with shrivelled roots or too much leaf. Anybody can try this experiment. The researchers found that the yields can be increased 20 to 30% just by cultivating at the right time.

Transplanting and Sowing Medicinal Plants according to the Moon Rhythm

Related Elements	Sowing or Transplanting when the Moon is in the Zodiac Sign of:	Part of the plants to be harnessed
fire	Aries Leo Sagittarius	fruits (also grains and cereals)
earth	Taurus Virgo Capricorn	roots
water	Cancer Scorpio Pisces	leaves
air	Libra Gemini Aquarius	flowers

Re-presenting Traditional Knowledge

When I was working in Peru, in the Andes, at 4000 metres above sea level, the farmers there told me that they always plant their seeds according to the moon. Farmers all over the world observe these natural laws. I have observed this with farmers in Africa, America and Asia: they all have some knowledge about how to harness the position of the moon. But nowadays a lot of such traditional knowledge has been lost.

A farmer in Peru told me: "If we use fertilisers, the moon doesn't matter." In his simple words he clearly stated that if we use fertilisers, we cut off the cosmic influences. Another farmer in a village, at 4,000 metres altitude, told me: "We want to produce food that in itself has medicinal value."

There is a clearly palpable loss of traditional knowledge. Children are not trained in going back to the countryside and asking the elder people: 'What do you know about our environment?' Instead of learning from the old people about the medicinal plants, etc., they generally say: 'Oh, the old people don't know how to read, so they are ignorant'. That is the tragedy in which a lot of traditional knowledge is lost.

We can observe that the younger generation rejects the oral traditions because they clash with modern knowledge presented in writing. When I worked in the Andes, we tried to present traditional knowledge in a modern language. There is a psychological aspect to it. When a young person reads his grandfather's knowledge in a nicely-presented, logically-written paper, he considers it to be important. And if he sees it on the TV, then it is even more important!

If traditional knowledge is presented in a format that young people can understand, they re-evaluate it. So, we produced a magazine with a picture of the grandfather and an article about the elaborate cultivation system that he had developed. This magazine was presented to all the farmers. Only then did they have the courage to go to that farmer and ask him questions. He finally became their teacher. He could only become their teacher because his insightful tradition was validated by a written system of knowledge. There is no school for such knowledge. It lives with individual people - I call them the 'rural scientists'.

New Dimensions of Cultivating Medicinal Parts

"Everything in the cosmos comes together to bring about the presence of one Flower..."

Biodynamic agriculture has developed for the healing of the earth, farm land and agricultural landscapes. It recognises and reinforces the basic principles at work in nature. Bios means 'life' and dynamics refers to 'energy'. Hence biodynamic agriculture suggests working with the energies which create and maintain life. The following excerpts are based on Mr. Caldas' lectures presented during a Colloquium on Biodynamic Cultivation that took place in Kalimpong in December 1996, and on his teachings imparted during visits over the past three years to ITTM. Part One of the Colloquium was published in AyurVijnana Vol. 5, autumn 1998. Part Two deals with the practical aspects of biodynamic cultivation.

To enhance the natural properties of medicinal plants we must observe nature and consciously help it evolve. Agriculture should normally bring about the fruitful combination of nature and culture. Biodynamic cultivation harnesses the life-giving forces of nature in the development of healthy and productive landscapes. Biodynamic principles were developed as a farming practice and as a scientific method by Rudolf Steiner at the beginning of this century. They are based on intimate wisdom gained by long observation of natural phenomena and their forces, as well as from spiritual research. Biodynamics seeks to harmonise plants and animals with their archetypes and farms with their landscapes and wider terrestrial and cosmic environments.

Drawing partly from traditional agricultural experience and wisdom, biodynamics, encompassing a wider potential than the organic method, respects cosmic rhythms and influences and weaves them into daily practical activity in order to call forth the subtle properties that are the essences of plants and animals. Thus, it is very suitable for the cultivation of medicinal plants, which depend on the fine properties of healing active principles, usually drawn naturally from the places where the plants grow.

The natural habitats of many Tibetan medicinal plants, for example, are the unique soils and clear starry nights of the stunning heights of the Himalayas and Tibetan plateau. Bringing these plants into cultivation poses a great challenge. Many formulations available through medical practitioners or sold through pharmacies possess only a fraction of their potential medicinal properties due to the methods used in commercial cultivation. On the other hand, the existing trade of plants collected from their wild original habitat poses the dangerous threat of extinction and permanent damage to those fragile mountain ecosystems. It has been with these challenges in mind that ECOTROPIC has started cooperating with ITTM's projects.

Together, our aim is to develop a cultivation system for medicinal plants which will on the one hand ease the need for wild collections and on the other hand preserve as much as possible of the medicinal and healing value of cultivated plants.

In the last article we dealt with the primary aspects of biodynamics. Now we would like to focus our attention on some of the important principles related to the management of soil fertility.

Soil Fertility of Plant Nutrition

The conventional approach to the fertilisation of soils has reduced that activity to a very simplistic and mechanical formula: plant ashes = 'x' amount of nitrogen; phosphorus potash = chemical fertiliser to be applied to the plant. The soil is considered a support substrate, and the nutrition of the plant plus water is provided directly to its roots. This has been at the basis of agricultural practices and productivity since the end of the last century. Yields have been raised tremendously, but mostly at the expense of the natural capital accumulated in the soil. Now, worldwide productivity faces an accentuated decline, which is not deterred by doubling the doses of chemical fertilisers given to the soil. This is a symptom of the reductionist approach that has permeated all aspects of life in this century. It has led to large scale destruction of landscapes and a decrease in the vitality and nourishing value of the food we eat. A clear indicator of this situation is the worldwide, multimillion dollar sales of complementary vitamins and minerals, now, at the turn of this century.

The Soil is Alive

In the biodynamic approach, we certainly use soil and plant analysis of nutrients as a reference for our practices, but that is the extent of its significance. We must understand the soil as part of the wider organism of the landscape and of the earth, possessing intrinsic life, rhythms and needs. The soil is not just a mineral substance but it is alive, it breathes, it has to be fed with adequate food and it can also become ill or malnourished. Certainly it can also die, when not given proper care. Obviously, then, ill soil cannot and will not produce healthy plants, able to maintain healthy human beings. It will produce weak plants that need a lot of 'remedies' to survive until they reach the consumer's plate. The same will happen with medicinal plants improperly cultivated, a fact which puts their use into question.

The question arises, how can we best feed the soil so that it will give us food and goods full of vitality as well as the life forces that are necessary for our physical, cultural and spiritual development?

By enriching the soil with life, life will be given to plants, and plants will then grow to their full potential. This will in turn benefit human beings and animals. Given the state of the earth, contaminated during the 20th century with chemical fertilisers and pesticides, this approach might be viewed as an appropriate remedy - a healing method of cultivation.

The Nature of Humus, the Humus of Nature

How can we simulate the soils found in the natural habitats of medicinal plants? What is the best food for the soil? We can answer this question by looking at how nature produces it, how it is 'cooked' in the forests and on the land. Plant and animal residues, moisture, and air are broken up by fungi, bacteria and small animals over a long period of time to produce the best food: humus - a full and balanced meal for the soil. Humus can be produced by composting. 'Compost' means exactly what the word indicates, a 'compound', a 'composite' of substances, a 'composition' of elements. It is very rich and has all the nutrients a plant wants: nitrogen, phosphorus, potash, calcium and sulphur, as well as over twenty trace elements. It consists of aminoacids, growth promoters, natural antibiotics, humic acids, and millions of beneficial bacteria, fungi, earthworm eggs, and so forth. Humus is the reserve food of millions of soil organisms. A soil with the appropriate amount and quality of humus enables plants to grow to their best potential. In short, humus is a concentrate of life forces, the basis for life, the sustenance of vegetation and thus for life on this planet, including human beings.

The strategy in the biodynamic system is not just to produce humus, but humus of the best quality, permeated with healing life forces. There is a difference between biodynamic and organic and other types of compost. It is like the difference between fermented milk and yoghurt or fine cheese. In addition, we must think in terms of the long term needs of the soil, not just the immediate needs of the cultivated plant. We work to build up the reserves of humus in the soil and live off the 'interest,' as we might do from our savings account. This means that the humus will be releasing nutrients to the plants as they are needed, in the right proportion, on demand, instead of the forced feeding done in chemical agriculture, where short term thinking exploits the natural reserves without giving them back.

Biodynamic Compost

As a complex substance, the production of good humus depends on the composition of materials that go into its production. We must consider the ingredients that go into our compost heap and the environment that is needed to break them down. For a proper process to develop, we must have the right carbon to nitrogen ratio. The coarser or more woody a plant is, the more carbon it will contain in relation to nitrogen, while the fresher and softer a plant is, the less carbon it will contain and the more nitrogen. This ratio is important because the bacteria and fungi that will decompose the compost need sufficient nitrogen and water content in order to work. Therefore, we must balance plants that have a higher content of fibre with plants, such as legumes, that are rich in protein or nitrogen. Or, we can also add nitrogen-richer materials, such as the residues of animals.

In biodynamic compost, cow manure has a special place. Cow-dung is a very balanced substance, containing the fibre that the cow has eaten, plus many nutrients, in an almost soluble form. The most important factor is the millions of microorganisms from the digestive system of the cow that bring life to the compost and continue the decomposition and breakdown of cellulose. When placed into a pile of plant material with proper moisture, the fibrous substances are digested the same way as in the intestine of a cow.

In the process of cooking food, it is heat that breaks down the ingredients so that they are easily assimilated. A compost that is properly built up should generate heat within the first few days. Fungi and afterwards bacteria will start to break the carbon bonds and thus release its stored energy for their activities. As this multiplies, a chain reaction takes place, and the light and energy that the plant has absorbed are released. A cold compost means that something is wrong, in the same way that a cooking fire stops if not enough air is allowed to penetrate to the stove. Thus heat is an important indicator that the process is developing in the right direction, that the compost has started 'to cook,' that this new digestive organ is alive. So, we must get the right balance of air, water, plant and animal materials and even wood ashes and minerals from the earth element. As with cooking, we must develop the right sense for the ingredients, the amount of water, the length of the process, the amount of heat required, and especially the amount of salt and spices. There is no other way to become a good cook than by doing it practically. Recipe books alone can never turn a person into a good cook. Only practice - tasting, smelling, and learning with mistakes - can accomplish this transformation.

Compost-making is an art which improves with time, and in the process, results in improved 'cooked' food for the soil. The more experienced the farmer and gardener, the better the food that is produced for the soil, and the better response we will get in terms of healthy nutritious plants and animals. Moreover, we will be able to cook special foods for different types of soils and plant requirements. Thus, we may recreate the best conditions for growth, also for our medicinal plants.

A good way of making a heap is to create a basis of woody material which can allow some air to be stored, followed by a layer of 30 cm of chopped green material, weeds, etc., sprinkled with wood ashes and rock phosphate, and made wet by a 'sauce' of cow- dung slurry. That mixture is then covered with a thin layer of cow-dung from the stable, mixed already in cow urine and waste fodder. That process starts again with a layer of 'chopped green vegetables, salt, spices and sauce', and so on, until the heap is about 1.30 mtrs. high and covered either with soil or mulch. Do not tread on the heap. If it becomes compact, make some holes in the heap with a stick so air can come in. The same compost can be prepared in boxes.

For the ingredients to cook, moisture must be available as well as heat. To measure the heat, leave a stick inserted and take it out to check the development. When the temperature in the heap decreases, earthworms invade it from the soil, and finish the job done by the microorganisms. They really love the compost heap! During their chewing, they mix the earth with plant material, which in turn produces calcium in their digestive system. The activity of earthworms in the soil is thus increased, further enhancing the healing of the soil and aerating it. The indicator of a good compost is that after 3 or 4 months, one cannot identify what has been put in. It has become a uniform, dark brown substance that has a sweet smell. A compost that is suitable for your area is developed by using the materials which are found locally.

After finishing the heap, one set of the biodynamic compost preparations is inserted. These are made from specially prepared herbs that will guide fermentation to achieve the best quality humus. These inoculators are made from yarrow (#502), camomile (#503), stinging nettle (#504), oak bark (#505), dandelion (#506) and valerian (#507). They permeate the compost with their qualities and processes related to specific nutrients. The resulting humus will carry these qualities right into the plant and help them in their development and metabolism. We hope to be able to produce these biodynamic inoculators at ITTM in the near future. We are already cultivating most of the herbs for this process.

The Plant between the Sun and the Earth

There are also two biodynamic preparations used for balancing the plant's growth between the sun and the earth. One preparation is applied to the soil, helping the plants develop deeper roots and activating the microorganisms in the soil. The other helps the plants improve light absorption, ripening and ultimately reconnecting to finer cosmic forces, which are important for the plants' medicinal and nutritive values. Silica (BD501) is applied to the part of the plant you wish to harvest when it is forming. The other (BD500), a humic compound, is applied to the soil. The first is applied very early in the morning, at dawn when the sun is rising, or 'breathing out,' which coincides with the rising of the plant's sap towards the tips. The latter, BD500, is applied when the sun is setting, or 'breathing in,' and the plant's sap and activity is drawing back to its roots. Thus, they are connected to the rhythm of the day. In fact, these two preparations are also made in connection with the rhythm of the year: Silica (BD501) is produced in the summer, and Humus (BD500) in the winter.

BD500 should be stirred by hand or with a stick in 20 litres of water, creating vortexes, clockwise and counter-clockwise for one hour.

This chaos, while reversing the movement, opens the water to the cosmos and creates a negative space receptive to environmental influences. It simulates the movement of water and air within powerful nature phenomena such as cyclones and whirlwinds, permeating the water with the qualities of the preparations. After the process, each drop then will radiate its influence over the soil and plant. Though machines are available, when done in a group of people, the stirring can also be looked at as an exercise for developing will and concentration, even meditation. The Silica preparation (BD501) is stirred in the same fashion but sprayed as a fine mist on the foliage in the early morning. The process to which they are subjected and the way of stirring results in a product possessing a concentration similar to a homeopathic remedy, which is required in very small amounts. So, 50 grams of BD500 and just 2 grams of BD501 are sufficient for one acre.

These preparations help adjust the environment, making it more suitable for the plant. If, for instance, one does not have the ecologically optimal situation for the cultivation of a medicinal plant, these preparations are helpful in adjusting a plant's ecological needs by balancing them with more or less of the required environmental conditions, thus coming closer to its natural habitat. Silica increases the absorption of light, compensating for poor exposure to the sun in the valleys, while more BD500 and humus simulates the valleys and shaded places and enhance leaf production. For medicinal plants that require root development, conditions will vary from those where leaf is the main medicinal component. The same applies to the balance between bitter and sweet substances, oils and other active principles. Of course the appropriate location of plants, regarding shade, exposure and types of soil, has to be determined, in conjunction with the above. Optimising conditions is achieved by the adequate use of biodynamic preparations and also specific compost types. The medicinal plant research project at ITTM will focus on such findings for Tibetan and other indigenous medicinal plants of this Himalayan region.

Pests as Indicators of Imbalances in Plant Development

When the soil is healthy and fertile, so are the plants, and, generally, few if any pests will appear. Modern scientific specialisation has failed to address this relationship between soil management and a particular problem of the soil. Instead of nitrogen fertilisers, which lead to aphids and fungus, or pesticides and fungicides, which only temporarily treat these symptoms, biodynamics recognises the cause and looks at the way the plant is feeding in order to balance its composition. A wide variety of plant extracts are known to discourage pests as well as providing strength to a plant. For instance, a herbal brew of Artemesia repels many pests. Nettles are especially good for aphids, and Equisetum arvensis acts as a natural mild fungicide. When we do not resort to toxic chemicals for the control of pests, natural predator insects, such as ladybird beetles, green lacewings and spiders, appear. They are the indicators that the ecosystem is in balance, and they help us to reduce pest problems. Many common herbs and medicinal plants can be used in this process. Cow urine can also be used, mixed with herbal extracts, to strengthen the plant, repel pests and reduce fungus problems.

Most Weeds are Medicines for Soil and Humans

The weeds around us reveal the health of the soil and at the same time improve its health. Weed-cover becomes more diverse as soil becomes richer and more porous. It is a symbiotic process where co-evolution takes place. Hard, dense soils are colonised naturally by hard grasses possessing powerful fibrous roots that slowly open it up, creating the environment for other weeds to develop. A variety of weeds can be established for the fine-tuning work of replenishing the soil's nutrients and other aspects important for its fertility. Instead of using herbicides, we can help by simply cutting the weeds and leaving them as mulch to recycle the absorbed elements back into the soil. By covering the bare earth with them, excess weeds will also be balanced, evaporation reduced and erosion prevented.

Conclusion

We have tried to briefly describe some of the practices and principles of the biodynamic method in these two articles. Still, we have barely scratched the surface of this complex agricultural practice and its understanding. In future issues of Ayur Vijnana we will continue outlining and deepening our knowledge of biodynamic cultivation of medicinal plants.

5

New Dinosaur Species Found in India

A reconstruction of the dinosaur's skull may shed light on how the continents drifted into their present positions and what might have caused the extinction of the dinosaurs.

When paleontologists Paul Sereno and Jeff Wilson arrived in India in 2001 to study a mixed collection of dinosaur bones gathered by Indian scientists 18 years earlier, they found the bones spread out on an office floor.

Sifting through the collection, they separated out the bones of a theropod, or meat-eating dinosaur. When they found the centre part of a skull, they recognized a horn resembling those of dinosaurs found in Madagascar. Their search continued, yielding a left hip, then a right hip, then a sacrum.

Sereno and Wilson consulted detailed, hand-drawn maps drafted by their Indian counterparts and discovered the bones had been buried next to each other, as if they had been connected.

"There was a Eureka! moment when we realized we had a partial skeleton of an undiscovered species," said Sereno, a paleontology professor at the University of Chicago and a National Geographic Explorer-in-Residence.

65-Million-Year-Old Bones

The bones were collected in 1983 by Suresh Srivastava of the Geological Survey of India (GSI) and Ashok Sahni, a paleontologist at Panjab University, during a search for dinosaur eggs and nests. Srivastava drew a detailed map to document the position of the fossil bones as they lay in the field. The scientists then stored the 65-million-year-old bones at a GSI office, where they stayed until Sereno and Wilson arrived.

Working with Indian experts, Wilson and Sereno reconstructed the skull of the new species, a stocky, 30-foot-long (9-meter-long) carnivore named *Rajasaurus narmadensis,* which means "regal dinosaur from the Narmada," the river region in western India where the bones were found. The project was supported in part by the National Geographic Society.

"We knew of fragments and bones [in India]," said Sereno, who has discovered new dinosaur species on five continents. "But this skull reconstruction offers the first glimpse into the lost world of the Indian dinosaur."

There were already two Jurassic dinosaur skeletons mounted in India. Neither, however, represents a single skeleton, but is instead based on composites of isolated bones.

"We know that there were carnivorous and herbivorous dinosaurs in India through individual bones, but we really don't know just how they looked because no two bones can be reliably said to belong to one individual," said Wilson, who is of the University of Michigan. "*Rajasaurus* is important because it represents a partial skeleton and preserves many details that clue us into its evolutionary relationships."

The reconstructed skull is missing some parts, but it has the most important pieces: the jaws and the brain case. Between 25 and 30 feet (7.6 and 9 meters) long, the *Rajasaurus* was heavy and strong, and walked on two legs.

"There are several anatomical details that make *Rajasaurus* a new species," said Wilson. "Perhaps the most striking is the horn it bears on its head. The horn was probably rather subtle. It may have been low and rounded."

A Carnivorous Family

The carnivorous *Rajasaurus*, which lived in the Cretaceous Period at the end of the dinosaur age, preyed on long-necked titanosaur sauropods, herbivorous dinosaurs that also roamed the Narmada region. Bones from both dinosaurs were found together.

Indian paleontologists recently found coprolites (fossilized dung) that provide additional clues to the diet of those titanosaurs.

"Large theropod eggs have also been described by our group from the area where the skeleton of *Rajasaurus* has been recovered, but it's difficult to relate the theropod eggs specifically to *Rajasaurus,*" said Sahni.

The scientists believe the *Rajasaurus* is related to a family of large carnivorous dinosaurs, most of which had horns, that roamed the southern hemisphere land masses of present Madagascar, Africa, and South America.

"People don't realize dinosaurs are the only large-bodied animal that lived, evolved, and died at a time when all continents were united," said Sereno.

Continental Drift

The *Rajasaurus* was likely like its contemporary, *Tyrannosaurus rex*—one of the last species to live before a catastrophe occurred some 65 million years ago.

What exactly caused the death of the dinosaurs is a matter of great debate. But the burial site of the bones found in India could provide a clue.

"The sediments in which these dinosaurs were found are closely associated with one of the biggest volcanic activities recorded in the last 500 million years on planet Earth," said Sahni. As much as a third of India's land mass is covered with lava. That makes it particularly difficult to find dinosaur bones. The task is made even more difficult because India is so heavily populated.

"The best place for a paleontologist to work is an empty desert," said Sereno.

Dinosaur discoveries could also help researchers improve their relatively scant knowledge of how India separated from Africa, Madagascar, Australia, and Antarctica and later impacted Asia to form the Himalaya.

"The details of the timing and route of India's northward migration are not well resolved," said Wilson. "Dinosaurs are good organisms to study the effects of continental drift because they cover continent-scale distances. That is, they seem to record connections between continents fairly well."

The new species will be described in the August issue of the *Contributions of the Museum of Paleontology* of the University of Michigan. Casts of the dinosaur's skull will be donated by the team to Panjab University and the Geological Survey of India.

Spices and Condiments—Nutmeg

Nutmeg is native of Molucces Island (Indonesia) and was introduced to India towards the end of the 18th century and is grown now in certain pockets of Kerala, Tamil Nadu, and Karnataka. They are important to confectionery, culinary and pharmaceutical industries. Nutmeg and mace also yield 7 to 16 and 4 to 15% of oil respectively. This oil is used for flavouring food products and liquors and also in perfumery industries. Oleoresin of nutmeg and mace is used almost entirely in the flavouring of processed foods. The nutmeg butter prepared by the expression of fat from the ground, cooked nutmeg is mostly used to impart spicy odour to the perfumes. The pericarp is used for making jams, jellies and pickles.

Botanical name of nutmeg is *Myristica fragrans* Houtt which belongs to family Myristicaceae. Nutmeg is a densely foliated evergreen tree growing to a height of 20m and above. The fruit is a fleshy drupe, spherical or slightly ovoid in shape and pale yellow in colour with a longitudinal groove in the centre.

Climate and Soil: Nutmeg thrives well in warm humid conditions in locations with an annual rainfall of 150cm and more. It grows well from sea level upto about 1300m above MSL. Areas with soils of clay loam, sandy loam and red lateritic are ideal for its growth. Both dry climate and water logged conditions are not good for nutmeg. The soil and climate prevailing in the hill slopes of Western Ghats and Eastern Ghats (from 700 to 1300m) are ideally suited to grow nutmeg.

Varieties: There is no named variety of nutmegs in India.

Propagation and Nursery Techniques: Nutmeg trees are usually propagated through seed. Large sized tree burst fruits from high yielding trees are harvested or collected as and when they fall down. The fleshy rind of the fruit as well as the mace covering the seed is removed before sowing. The seeds should be sown in specially prepared nursery beds immediately after collection. If there is delay in sowing, the seeds may be kept in trays filled with moist sand or stored in ploybags or moss, which keep them viable for 15 days.

Regular watering is necessary for good germination. The germination may commence from about the 30th day and lasts upto 90 days after sowing. About 20 days old sprouts are transplanted to polythene bags containing a mixture of good soil, sand and well decomposed cowdung (3:3:1). About 18 to 24 months old seedlings are used for transplanting in the field. Air layering, inarching and budding have also been attempted with varying degree of success, but the epicotyl grafting technique developed by NRCS, Calicur is found to be promising now.

Planting: Seedlings are transplanted in the main field when they are 12 to 18 months old. Recommended spacing is 8 x 8m. Pits of 60cm cube are dug and filled with compost and topsoil. Seedlings are planted carefully in the centre of the pits during rainy seasons. Young plants are provided with artificial shade and irrigated during summer months. The basins are mulched with locally available trashes. As nutmeg requires light shade especially at early stage, fast growing shade plants like *Erythrina indicia*, *Elithosperma* or *Gliricidia maculate* or bananas are planted in between them a few months prior to planting. As nutmeg plants grow, these shade plants may be thinned out. Through not essential, irrigation is advantageous to grown up trees for getting better yield.

Nutmeg can be advantageously grown mixed with coconut and arecanut. When coconut is planted at a spacing of not less than 8 x 8m one plantations raised with a spacing of 2.7m x 2.7m, nutmeg can be planted at every third row of arecanut so that within the square formed by four nutmeg plants there will be nine arecanut seedlings. In coffee based mixed system also, nutmeg can be planted at a spacing of 7-8m.

Manures and Fertilizers: Nutmeg requires heavy manuring for proper growth and yield. Farm yard manure or compost at the rate of 10kg per plant may be applied during the first year of planting. The quantity should be increased every year so that a well-grown tree of about fifteen years and above may receive 50kg organic manure. During the first year, fertilizers to supply 20g nitrogen, 18g phosphorus and 50g potash are applied per plant. The fertilizer doses may be increased gradually every year till a well grown tree of fifteen years or more receives about 500g nitrogen, 250g Phosphorus and 1000g potash annually. The fertilizers are applied in two equal split doses, first in May-June along with organic manure and the other in September-October. Shallow trench is dug around each tree at a radius of 1 to 1.5m away from the trunk for applying manures and fertilizers and covered after application of manures. As nutmeg has a very shallow root system, trenches should not be deep and no intercultivation is usually done very close to the trunk. However weeding is done periodically.

Harvesting: Seedling trees start bearing in 7 to 8 years while grafts start bearing in 4 to 5 years. They attain full bearing stage after 15 to 20 years and may yield upto 60 years. Nutmeg trees flower throughout the year with a peak in certain months. Hence, though fruits are seen throughout the year, the peak season of harvest is from June to August. The fruits take 9 months from flowering to harvest. When the fleshy rind of the nut splits open, the fruits are fully ripe for picking. They are either plucked from the tree or are allowed to drop on to the ground and then collected. After removal of the outer fleshy rind, mace is detached from the seed shell by hand and flattened out. It is then allowed to dry slowly in the sun for ten to fifteen days. During drying the mace gradually becomes brittle and horny and attains a yellowish brown colour. The seeds are dried separately for 4 to 8 weeks either in the sun or in artificial heat until the kernel rattles inside the shell, which is then broken open with a wooden mallet, and the kernel is taken out. The fleshy pericarp can be used for making pickles, jams and jellies.

Yield: Individual nutmeg fruit weights on an average 60g of which the seed weighs 6 to 7g, mace 3 to 4g and the rest pericarp. A fully-grown tree under normal conditions may produce on an average 2,000 to 3,000 fruits per year. On the assumption that about 65% of the trees in a garden are female and are optimum yields, about 800 kg nutmeg and 100kg mace can be expected per hectare.

Plant Protection

Disease	Symptoms	Control measure
Die back (Diplodianatalensis)	Drying up of mature and immature branches from tip downwards.	Remove the infected branches; smear Bordeaux paste in the cut end of the branches.
Thread blight (Marasmius sp.)	Blighting of the affected stems or leaves	Spray 1.0-% Bordeaux mixture.
Fruit rot (Diplodia natalensis and phytophthora sp.)	Splitting of immature fruits and subsequent rotting of mace.	Spray 1.0-% Bordeaux mixture when the fruits are half-mature.

Coriander

Coriander *(Coriandrum sativum L.)* is an annual herb, mainly cultivated for its fruits as well as for the tender green leaves. It is native of the Mediterranean region. In India, it is grown in Andhra Pradesh, Tamil Nadu, Karnataka, Rajasthan and Madhya Pradesh. Major portion is though consumed locally; a small quantity is being exported now. The fruits have a fragrant odour and pleasant aromatic taste. The odour and taste are due to the essential oil content, which varies from 0.1 to 1.0 % in the dry seeds. These essential oils are used for flavouring liquors, coca preparations in confectionary and also to mask the offensive odours in pharmaceutical preparations. The dried ground fruits are the major ingredients of the curry powder. The whole fruits are also used to flavour foods like pickles, sauces and confectionary. The young plants as well as the leaves are used in the preparation of chutney and are also used as seasonings in curries, soups, sauces and chutneys. It has medicinal properties too. Fruits are said to have carminative, diuretic, tonic, stomachic and aphrodisiac properties.

Coriander belongs to the family Apiaceae. It is a smooth, erect annual herb 30 to 70 cm high, lower leaves broad with crenately lobed margins, upper leaves finely cut with lineary lobes, flowers small, white or pink in compound terminal umbels, fruits – schizocarp, globular, yellow-brown, ribbed, 2 seeds, ripe seeds are aromatic.

Climate and Soil: It is a tropical crop and can be grown throughout the year (except very hot season *i.e.* March-May) for leaf purpose, but for higher grain yield it has to be grown in specific season. A dry and cold weather free from frost especially during flowering and fruit setting stage favours good grain production. Cloudy weather during flowering and fruiting stage favours pest and disease incidences. Heavy rain affects the crop. As an irrigated crop, it can be cultivated on almost all types of soils provided sufficient organic matter is applied. Black cotton soils with high retentivity of moisture is best under rainfed conditions.

Varieties: Many improved varieties of coriander are now available for cultivation in Tamil Nadu, Andhra Pradesh, Gujarat and Rajasthan states.

Variety	Parentage	Characteristics
CO1	A pure line selection	Released by TANU, Coimbatore. Tall plant, many umbels per plant, suitable for green and grains. Duration 110 days. Yield 500kg per ha.
CO2	A reselection from culture P_2 of Gujarat	Released by TANU, Coimbatore. High yield, dual purpose variety, tolerant to drought, oil 0.3%. Duration 90-110 days. Yield 600-700kg per ha.
CO3	Reselection from Acc. No. 695	Released by TANU, Coimbatore. High yield, dual purpose, medium size grain, seed oil 0.38-0.41%. Duration 103 days. Yield 640kg per ha.
Gujarat Corinader-1	A selection from local	Released by GAU, Jagudan. High yield, more number of branches, seeds bolder and greenish in colour. Duration 112 days. Yield 1100kg per ha.
Gujarat Coriander-2	A selection from CO2	Released by GAU, Jagudan. High yield, more branches, dense, foliage, umbels large size, grain purpose variety, bold seeds, no lodging. Duration 110-115 days. Yield 1500kg per ha.
Rajendra Swati	A mass selection from germplasm type	Released by RAU, Dholi. High yield potential, suitable for intercropping, fine seeded, rich in essential oil, resistant to stem gall disease. Duration 110 days. Yield 1200-1400kg per ha.
Rcr-41	Recurrent selection from UD 41	Released by RAU, Jobner. High yield, tall erect, suitable for irrigated areas, resistant to stem gall. Duration 130-140 days. Yield 1200kg per ha.
Swati	Mass selection	Released by APAU, Guntur. High yield, semi erect, suitable for delayed sowing. Duration 80-90 days. Yield 885kg per ha.
Sadhana	Mass selection	Released by APAU, Guntur. High yield, suitable for rainfed areas, semi erect, resistant to aphid and mites. Duration 95-105 days. Yield 1000kg per ha.

Land preparation: For raising a rainfed crop, the land is ploughed 3 to 4 times following rains and field must be planted immediately to break the clods and to avoid soil moisture. For irrigated crop the land is ploughed twice or thrice and beds and channels are formed.

Sowing: In the North and Central parts of India and Andhra Pradesh, it is mostly grown as a Rabi season crop and hence sowing is done between middle of October and middle of November. In certain pockets of the above area, late kharif crop is sometimes sown in August-September. In Tamil Nadu, as an irrigated crop, coriander is raised in June-July and September-October. In the first season, the crop matures late with an extended growth phase during January-February. The growth and the yield rainfed conditions, it is sown during September-October, at the onset of Northeast monsoon and harvested during January-February.

A seed rate of 10 to 15 kg per hectare is required. Seeds stored for 15 to 30 days record better and early germination than freshly harvested seeds. Seeds soaked in water for 12 to 24 hours before sowing also enhances better germination. The seeds are split into two halves by rubbing and generally done in rows spaced at 30 to 40 cm apart with 15 cm between hills. Soil depth should not exceed 3.0 cm. Three to five seeds are sown in seeds are broadcast and covered with country plough. Germination takes place in 10 to 15 days.

Manures and Fertilisers: About 10 tonnes of farmyard manure is applied at the time of last preparation. In addition, the following fertilizers may be applied.

Name	N		P	K
	Basal	30 days after sowing	Basal	Basal
Irrigated crop	15	15	40	20
Rainfed crop	20	-	30	20

Irrigation: First irrigation is given 3 days after sowing and thereafter at 10 to 15 days interval depending upon the soil moisture available in the soil.

Interculturing: The first hoeing and weeding and weeding are given in about 30 days. Thinning the plants is also attended simultaneously, leaving only two plants per hill. Depending upon the growth one or two more weeding are done.

Harvesting and Yield: The crop will be ready for harvest in about 90 to 110 days depending upon the varieties and growing season. Harvesting has to be done when the fruits are fully ripe and start changing from green to brown colour. Delaying of the harvest should be avoided lest shattering during harvest and splitting of the fruits in subsequent processing operations. The plants are cut or pulled and poled into small stacks in the field to beating with sticks or rubbing with hands. The produce is winnowed, cleaned and dried in partial shade. After drying, the produce is stored in gunny bags lined with paper. The rainfed crop yields on an average 400 to 500 kg/ha and the irrigated crop 600 to 1200kg/ha.

Plant Protection: At the seedling stage coriander is often attacked by the leaf eating caterpillars and semi-loopers and at the flowering stage by the aphids. Spraying the crop with methyl demeton (0.05%) is recommended to control the aphids but a flowering stage the use of any insecticide would kill the bee population affecting pollination in the crop.

Powdery mildew (*Erysiphe polygoni*) is a serious disease, which ruin the crop if allowed unchecked in the initial stage itself. Spraying wettable sulphur 0.25% or 0.2% solution of Karathane twice at 10 to 15 days interval is recommended. Grain mould is caused by *Helminthosporium* sp., *Alternaria* sp., *Carvularia* sp. and *Fusarium* sp. It can be controlled by spraying carbendazim 0.1% 20 days after grain set.

Garlic

Garlic, a native of Southern Europe is one of the important bulb crops grown and used as a spice or condiment throughout India. Gujarat followed by Orissa are the largest producing states. It possesses a high natritive value, its preparations are administered as a cure against stomach disease, sore eyes and ear ache. It is commonly used in the preparation of various dishes. Allicin, the principle amoebic dysentry and is also having many other medicinal properties.

Climate and Soil: It is grown under a wide range of climatic conditions. However, it cannot stand too hot or too cold weather. It prefers moderate temperature in summer as well as in winter.

Short days are very favourable for the formation of bulbs. It can be grown well at elevations of 1000 to 1300 m above MSL. Garlic requires well drained loamy soils, rich in humus, with fairly good content of potash. The crop raised on sandy or loose soil does soils, the bulbs produced are deformed and during harvesting, many bulbs are broken and bruised and so they do not keep well in storage.

Varieties: There is no distinct variety of garlic. Local varieties are either white in colour and have fairly big bulbs with a better keeping quality and a higher yield or red in colour with pungency. Tamil Nadu Agricultural University has recently released one improved variety by clonal selection *viz.*, Ooty 1 Garlic. It is a high yielder (17t/ha) with a shorter duration of 120 to 130 days. The bulbs are big sized weighing 20 to 30g and each bulb has 22 to 25 cloves, which are dull white in colour.

Propagation: Garlic is propagated by cloves. All the cloves are planted except the long slender once in the centre of the bulb. Bulbs with side growth should be discarded. Healthy cloves or bulbils free from disease and injuries should be used for sowing and about 150 to 200 kg cloves are required to plant one hectare. They are sown by dibbling or furrow planting.

- *Dibbling:* The field is divided into small plots convenient for irrigation Cloves may be dibbled 5 to 7.5cm deep, keeping their growing ends upwards. They are laced 7.5cm apart from each other in rows of 15cm apart and then they are covered with loose soil. June-July and October-November are the normal planting seasons for garlic.
- *Furrow Planting:* The furrows are made 15 cm with hand how or a cotton drill. In these furrows, cloves are dropped by hand 7.5 to 10 cm apart. They are covered lightly with loose soil and a light irrigation is given.

Manures and Fertilisers: About 25 tonnes of farm yard manure is applied as a basal dose along with 60kg Nitrogen and 50 kg in each of Phosphorus and Potash. Forty five days after planting 60kg Nitrogen is applied again as top dressing.

Irrigation: First irrigation is given after sowing and then field is irrigated every 10 to 15 days depending upon the soil moisture availability. There should not be any scarcity of moisture in the growing season, otherwise, the development of the bulbs will be affected. The last irrigation should be given 2 to 3 before harvesting for making it easy without damaging the bulbs. In South India hills, they are mostly grown as a rainfed crop.

Intercultural Operations: First interculture is given with hand hoe one month after sowing. Second weeding is given one month after the first (about two and half months from sowing) loosens the soil and helps in the setting of bigger and well filled bulbs. The crop should not be weeding out or hoed at a later stage because this may damage the stem and impair the keeping quality.

Harvesting: Garlic is a crop of 4 ½ to 5 months duration. When the leaves start turning yellowish or brownish and show signs of drying up, the crop is ready for harvest. The plants are then pulled out or uprooted with a country plough and are tied into small bundles which are then kept in the field or in the shade or 2-3 days for curing and drying so that the bulbs become hard and their keeping quality is improved. The bulbs may be stored by hanging them on bamboo sticks or by keeping them on dry sand on the market, the dried stalks are removed and bulbs are cleaned. Well cured garlic bulbs can be kept for 1 to 1 ½ months in an ordinary well ventilated room. If dust smoke is given to it, the bulbs can be stored for 8 to 10 months. They can also be stored at 32^{0}F with 60% R.H. Average yield level is 6 to 8 t/ha.

Plant Protection: Thrips cause withering of the leaves. Application of methyl demeton 25EC 1 ml/litre will check the incidence. Leaf spot is the most important disease. Spraying Dithane M-45 at fortnightly intervals at 2.5g in one litre of water is recommended.

Curry Leaf

Curry leaf plays an important role as a condiment in the culinary preparation of South Indian dishes. It is cultivated in field scale in Coimbatore, Periyar, Madurai, Salem and Trichy districts of Tamil Nadu and in Dharward, Belgaum and Uttara Kannada of Karnataka State. The botanical name of the curry leaf is *Murrya koenigii* Spreng. It belongs to the family Rutaceae.

Climate and Soil: It does not require a specific climate and can come up in dry climate too. In places where minimum temperature goes below 13^{0}C, the growth of shoot will be slightly affected. Though it can be cultivated in most of the soil types, it comes up well in light textured red soils.

Varieties: Farmer prefers local varieties, which have pink midrib. University of Agricultural Science, Dharwad has recently released two cultivars viz DWD – 1 and DWD –2 which have an oil content of 5.22 and 4.09% respectively. Both the varieties are having strong aroma.

Preparation of Field and Planting: Mainfield should be ploughed repeatedly. Normally, a spacing of 90 to 120cm is followed on either side. One month before planting, pits of 30 x 30 x 30cm dimensions are dug out and are filled with top soil mixed with well decomposed farm yard manure at the time of planting. Healthy seedlings are planted in the centre of the pits. Then long furrows are formed connecting all the pits to facilitate easy irrigation.

The seedlings are irrigated once in five to seven days upto three years and once in 15 days afterwards. The field should be kept free of weeds. Normally curry leaf plants are not fertilized with inorganic source by the farmers. However, for a better growth and yield, each plant may be fertilized with 20kg of Farmyard manure besides 150g nitrogen, 25g phosphorus, and 50g of potash per year.

Harvesting and Yield: At the end of first year, the crop comes to first harvest. The expected yield from one hectare is given below:

Age of plantation	Yield of leaves (kg/ha)
First year	400
Second & third year	2000 to 2200
Fourth year	2500
Fifth year onwards	3500 to 5000

The plantation can be kept for 20 to 25 years depending upon the management practices followed.

Plant Protection: Aphids attack the plant when the plants are in vegetative stage and spraying of dimethoate at the rate of 2ml/litre of water is recommended. Leaves from such sprayed plants should not be harvested for 10 days lest residual toxicity of the pesticide. Leaf spots sometimes occur and spraying carbendazim at the rate of one gram per litre is recommended.

Tamarind

Tamarind, *Tamarindus indicus* L (Family: Caesalpinaceae), is native to tropical Africa and is now widely planted and naturalised everywhere in the tropics. It grows widely in the tropical and subtropical regions of subcontinental and is also planted along avenues, in parks etc. The fruits are used for various culinary purposes all over India as they contain tartaric acid (8%). The seeds give a gum used for binding and the wood has a good timber value.

It is a medium sized tree with a short strong trunk. The leaves are pinnately compound with 10-20 pairs of small leaflets. The flowers are small, scented, and attractive with yellow and red coloured. Pods are flattened and very considerably in size and shape. When ripe, the fruits are stiff and brittle. The most valuable part of the tamarind fruit is the brown and sweetly acid pulp. There are reddish-pulp types, which are considered the best.

Varieties: Recently Tamil Nadu Agril. University (H.C.R.I, Periakulam) has released one improved tamarind variety PKM.1-Tamarind by clonal selection. It has a high pulp recovery of 39%, pobs are less fibrous and pulp is also very sweet. It yields 250kg from ninth year. Another local selection *viz.* Urigam, a long, podded type, measuring more than 20 cm in length are also becoming popular among the growers. The grafts of this selection also start bearing from third year onwards.

Climate and Soil: Tamarind can thrive in tropical and subtropical climate excepting in places experiencing frost. Similarly, it is not exacting in its soil requirement. It thrives even in sodic and saline soils, ravines and degraded land.

Propagation: It is generally propagated through seeds. Vegetative propagation through whip cum inarching method gives more than 90% success. Patch budding on 9 months old saplings also gives more than 90% success.

Planting: It may be planted from June to November in South India and a spacing of 10 x 10m may be adopted to accommodate 100 plants, per ha. A pit size of $1m^3$ may be dug for planting and is refilled with topsoil and farm yard manure for good establishment of the saplings.

Harvesting and Yield: The seedlings take 8 to 10 years for fruiting and the graft take 4 to 5 years for fruiting. It is not uncommon that trees with more than 100 years are yielding normally. There is a tendency of alternate bearing in this crop. Fruits are harvested from December to April. Though the graft starts yielding from fourth years, good yield of 250kg per tree is obtained from ninth year in PKM-1 variety.

Plant Protection: Among the pest, mealy bugs and scales are often causing serious concern especially in the nursery and can be controlled by spraying 0.1% monocrotophos. Powdery mildew sometimes infects the leaves and can be controlled by spraying 0.1% Karathane.

Allspice

The allspice of commerce is the dried immature fruit of the tree *Pimenta dioica* (Family: Myrtaceae). It is indigenous to West Indies. Jamaica is the main producer of allspice. Its flavour is said to resemble a blend of cinnamon, clove and nutmeg. There are a few trees available in the Mahendragiri hills of Negercoil area, Ambalavayal in Kerala and Kallar and Burliar of the Nilgiris.

It is a small evergreen tree, flowers are white and branch trichotomously in the axils of upper leaves. Flowers are structurally hermaphrodite but functionally dioecious. Stamens numerous, above 100 in barren trees and 50 in bearing ones. It flowers during March-June and matures in 3 to 4 months after flowering. Fruit is a two-seeded berry. Male trees flower earlier.

Cultivation: It is raised through seeds. The seeds can be stored as ripe berries after collection without extracting seeds upto three weeks. But viability gets reduced slowly after a period of three weeks and is lost completely after nine weeks. The seeds are sown in raised beds of 15 to 20cm high, one metre width and convenient length.

The beds may be made of loose soil-sand mixture having sand in the top layer. The seeds germinate in 15 days and continue upto 40 to 45 days. The seedlings can be transferred to bags 3 weeks after their emergence above ground level. Six months old seedlings are planted at a spacing of 6m either way. As there is no manurial schedule available for allspice, the schedule recommended for clove can be followed for this. Plants start flowering when they are 7 to 10 years old, but the peak harvest is obtained from 15 to 20 years.

Harvesting and Processing: The berries are harvested when fully grown, but still green. They are sun dried and are frequently turned down with ladles. The process of curing lasts 3 to 12 days, till the berries are completely dried. Curing is complete when the berries become crisp and produce a metallic sound if shaken. A tree yields 50 to 60 kg of dry berries. The Pimento berry oil is distilled commercially in Jamaica. The oil yield from berries varies from 3.3 to 4.5% and its eugenol content is 65 to 66%.

The pimento leaf oil is also distilled only in Jamaica from the leaves removed during the harvesting of the berries. Oil yield from dried leaves ranges from 0.7 to 2.9% while fresh leaves yield 0.35 to 1.25%. It is a flavouring agent in culinary preparations like soups, sauces, ketchup, pickles, meat, and curry powder and in bakery items. In liquor industry also, it is used as a favourite ingredient.

Vanilla

Vanilla is native of the Atlantic coast from Mexico to Brazil. It is grown on a plantation scale in Java, Mauritius, Madagascar, Tahiti, Seycheles, Zanzibar, Brazil and Jamaica and other islands of the West Indies. Malagassy Republic grows 70 to 80 per cent of the world's crop of Vanilla bean followed by Reunion. U.S.A is the largest importer. This spice was introduced to India as early as 1835. Its commercial cultivation is now restricted to Wynad of Kerala and Nilgiris of Tamil Nadu. Recently, the demand for natural vanilla is on the higher side.

It is an orchid, belonging to the family Orchidaceae. There are two important species of vanilla *viz. V. planifolia* and *V. pompana*. The former species produces short thick pods whereas the latter one has the largest pods. *V. planifotia* has opposite, sessile leave of 10 to 23 cm long which are oblong in shape.

Climate and Soil: Vanilla requires a warm climate with frequent rains and prefers an annual rainfall of 150-300 cm. Partially uncleared jungle lands are ideal for establishing vanilla plantations. In such locations, it would be necessary to retain the natural shade provided by lofty trees which allows penetration of sunlight to the ground level and to leave the soil or the rich humus layer on the top undisturbed. However, vanilla is cultivated in varied types of soils from sandy loam to laterites.

Propagation: The crop is usually established by planting shoot cuttings. If possible cuttings with 18 to 24 internodes should be used as they come to flower earlier than shorter cuttings. But the length of the cuttings are to be adjusted depending on the availability of the planting material and the area to be planted. However, cuttings with less than five to sic inernodes and shorter than 60 cm in length should be avoided when planted directly in the main field. For raising rooted cutting in polybag even two nodes cutting can be used. It can also be propagated by using tissue culture methods. However, care should be taken to use plantlets not less than 30 cm in length.

The leaves of the fourth to fifth nodes from the tip are removed and the cutting is kept loosely rolled up in a cool, shaded place for tow to three weeks. When ready for insertion, the cutting must be handled very carefully. The lower three to four internodes are placed in a shallow trench 3-4 cm deep and about 10cm wide. The evacuated soil is used to loosely fill this trench. This operation is usually carried out at the beginning of the rainy season.

Land Preparation: Preparing the soil for prospective pepper or vanilla plantations must take into account the need to supply each vine with a support or stake upon which it can climb. Later it will be seen that these supports are divided into two categories non-living and living. In the former, site preparation is unaffected because it is possible to put the non-living support, for example a wooden stake, in place at any time after the soil has been cultivated by general ploughing or hole preparation. Where living supports are used, these must be established before taking the cuttings from the pepper or vanilla plants. The supports most often used for pepper are either plants which are already in the plantation or trees from original forest growth, left in place during land clearance. In the latter case, the fact that the supports are already present makes it essential that the holes at the foot of each support are made by hand.

Planting: Vanilla, being a climbing vine requires support for growing. It flourishes well in partial shade of about 50 per cent sunlight and low branching trees with rough bark and small leaves are grown for this purpose. Some of the trees now being used *include Glyrideidai, Erythrina, Jatropha carcas, Plumeria alba* and *Casuarna equisetifolia*. If the support selected is a legume, it will be able to enrich the soil also. The growth of live standards is to be adjusted as to make them branch at a height of 120 to 150 cm, to facilitate training of the vines around the branching shoots. The standards are planted at a spacing of 2.5 to 3 metres between rows and two metres between rows and two metres within the row making a population of 1600 to 2000 trees per hectare. If limb cuttings are used for planting supports the ideal time is with the onset of rains after the summer, and it should be atleast six months before planting vanilla cuttings.

Vanilla is generally planted at a time when the weather is not too rainy or too dry. The months of August-September are ideal for vanilla cultivation. Cuttings for planting should be collected in advance, and after removing three or four basal leaves, dipped in one per cent Bordeaux mixture and kept in shade to loose moisture for about a week. Since establishment of cuttings is almost cent per cent, planting of single cutting per support is enough.

The defoliated part of the vine is laid on the loose soil surface and covered with a thin layer of about two to three cm soil. The basal tip of the cutting should be kept just above the soil to prevent rotting. The growing end is gently tied to the support for climbing by the aerial roots.

The cuttings are shaded with tall dry grass, palm fronds or with other suitable materials. In dry soil, a light sprinkling of water helps for early establishment of cuttings. It takes about four to eight weeks for the cuttings to strike roots and to show initial signs of growth. Vanilla can also be planted as an intercrop in coconut and arecanut plantations.

Mulching: This term refers to the act of spreading, on a cultivated plot, a fairly dense layer of a material which is usually, but not necessarily, of vegetable origin. This layer which should be as durable as possible, protects the soil from run-off and exposure to the sun, regulates rainfall infiltration, slows down evaporation, arrests or at least considerably restricts wed growth and is generally favourable to growth and yield since it also adds humus to the soil on decomposition. Mulching is used on many of the species studied, and is applied in various ways. mulching of vanilla is carried out as soon as possible, after planting. A mixture of grasses and leguminous species is recommended.

Training: If the vine is permitted to grow up on a tree, it will rarely blossom, so long as it is growing upward. Hence, vines are allowed to grow upto 1.50 m and then trained horizontally on the branch of supports and latercoiled round them. This induces more flower production in this portion of the vine.

Flowering: The vines commence flowering in the second or third year depending on the length of cuttings used due to the peculiar structure of the flower, artificial pollination by hand is the rule for fruit setting. The procedure involved is simple and done easily by children and women. Using a pointed bamboo splinter or pin, another is pressed against the stigma with the help of thumb and thus smearing the pollen over it. Generally 85 to 100 per cent success is obtained by hand pollination. The ideal time for pollination is between 6 a. m to 1 a.m. Unfertilized flowers fall within two or three days. Normally 5 to 6 flowers per inflorescence and a total of not more than 10 to 12 inflorescences per vine are pollinated. The excess flower buds are nipped off to permit the development of other pods. Pods take six weeks to attain full size from fertilization but takes 4 to 10 months to reach full maturity depending upon the locations.

Maintenance of Plantation: Once established, the vines have to be given constant attention. The plantation should be visited frequently to train the vines to grow at convenient level, to regulate the growth of the vines and the supports, to watch for disease and pests and to always keep leaf mulch around the vines. Any operation done in the plantation should not disturb the roots, which are mainly confined to the mulch and surface layer of soil. In vanilla plantations provided with living supports, adjusting the shading is linked with correct pruning of the supports, a task, which requires care and attention. In the first year, it is enough to prune the lateral branches so as to obtain a sufficiently high single trunk. Further growth in height is then prevented by topping the tree, which encourages the formation of a canopy but still provides light shade. Leucaena leucocephala is very well suited to this approach, as the pruning, which is left at the base of the tree, provide the soil with nitrogen-rich organic matter. With vanilla, the shading provided by the living support is often inadequate. It can be supplemented by planting a range of shade trees, for example, Albizzia lebbeck and Inga edulis.

When the support trees grow up, they are pruned early to induce branching. It is desirable to develop an umbrella shape for the trees to give better shade and protection to the growing vines. If the trees do not drop off leaves they are pruned before the commencement of heavy rains to allow in more sunlight. The pruned vegetation is chopped and applied as a mulch in the plantation. The way in which the vine is trained has an effect on flower production. If the vine is permitted to grown up on a tree it will rarely blossom so long as it is growing upward. For convenience of cultural operation the vines are allowed to grow up to a height of 1.2 to 1.5 metre and then trained horizontally on the branch of supports and later coiled round them. Alternatively two bamboo splits can be tied to two adjacent support trees and can be utilized for training the vines. Coiling of vines in this manner helps to accumulate carbohydrate and other flower forming materials, beyond the bend and to induce flower production in this portion of the vine.

Harvesting and Curing: When immature, the bean is dark green in colour, but when ripe yellowing commences from its distal end. This is the optimum time for harvesting the bean. If left on the vine the bean turns yellow on the remaining portion and starts splitting, giving out a small quantity of oil reddish brown in colour, called the Balsam of Vanilla. Eventually they become dry, brittle and finally become scentless. Therefore, the artificial methods are employed to cure vanilla. Vanillin is developed as a result of the enzyme action on the glucosides contained in the beans during the process of curing. Basically any curing method involves the following four stages.

1. Killing the vegetative life of the beans to allow the onset of enzymatic reaction
2. Raising temperature to promote this action and to achieve rapid drying to prevent harmful fermentation.
3. Slower drying for the development of different fragrant substances.
4. Conditioning the product by storing for a few months.
   i. *Peruvian Process:* Curing is done by hot water. In this process the pods are dipped in boiling water. The ends are tied and hung in the open. They are allowed

to dry for twenty days. Later they are coated with castor oil and afterwards tied up in bundles.

ii. *Guiana Process:* The pods are collected and dried in sun till they shrivel. Later they are wiped and rubbed with olive oil. The ends are tied up to prevent splitting and then bundled.

iii. *Mexican Process:* The harvested pods are kept under shade till they shrivel. Then they are subjected to sweating. This operation is carried out for two days depending on the weather conditions. In warm weather, the pods are spread over blankets and exposed to the sun. During midday the blanket is covered over and bundled and left in the open for the rest of the day. They are wrapped in blankets in the night to maintain continuous fermentation and sweating. The pods should be wrapped in blankets when they are hot to touch. This process is repeated for 7 to 12 days till they become dark brown in colour, soft and flexible. They are packed in tins and sealed. The Mexican process yields 4.15 to 4.40 per cent of vanillin content.

When the weather is cloudy, the pods are bundled in bales and wrapping with woolen cloth covered with banana leaves. They are subject to radiation of heat by maintaining the temperature of air-over at 50^{0}C for 24 hours. Thereafter, they are dried to change the colour. Then they are spread in dry place and finally packed and sent to the market.

The most desirable beans will be 18 to 25 cm long, dark brown. Highly aromatic, fleshy, free from mold, insects, and blemishes and somewhat oily in appearance. There are three grade *viz.* Grade-1, which includes whole beans of minimum 11 cm length, and grade-2 and 3 will have a minimum of 8 cm length. Vanillin extract is taken from the cured beans by hydroalcoholic extraction. The vanillin content of the properly cured beans will not be less than 2.5 percent.

Yield: The yield of vanilla varies depending upon the age of vines and the method of cultivation. Normally it starts yielding from the third year and the yield goes on increasing till the seventh or eighth year. Thereafter the yield slowly starts declining till the vines are replanted after another seven to ten years. In one acre you can plant about 1000 vanilla plants. Each plant is expected to yield about 500 grams of green beans per year. Under reasonable level of management, the yield range of a middle aged plantation will be about 500 kg of green beans per acre. If we calculate with minimum price of Rs. 150 for one kg, a year's income comes up to Rs. 75,000 per acre. Apart from the sale of green beans many vanilla farmers today are also making money by selling vanilla planting materials. Experts suggested that first time vanilla farmers should start by growing a few plants on a small plot of land. And then depending on the result they can expand the acreage.

Market for Vanilla: The main application of natural vanilla is for flavouring ice creams and soft drinks. It is estimated that nearly 300 tonnes of vanilla beans is used in U.S.A every year in the preparation of cola type drinks. The major industrial purchasers of vanilla are pharma companies and soft drink companies like Coke and Pepsi. However the fact remains that market for natural vanilla essence is today largely only confined to the West. There is no market in India at the moment for vanilla essence. The domestic market in India is restricted to the green vanilla beans. In India processing companies buy green vanilla beans from the farmers, process it and then export the same to foreign buyers who then do the extraction of vanilla essence.

World production of vanilla beans is approximately 3000 tonnes per annum. Madagascar provides about 50 per cent of the world supply and the rest is from Indonesia. Comoro and Reunion. Production in Indonesia is nearly 500 tonnes. The present international demand from vanilla is about 19,000 tonnes.

India has just come into the market for production. Our production last year was a meagre 30 tonnes only. India is still a very insignificant player in vanilla. It will take some more time before we make our presence felt in the world markets. Presently Indian farmers are getting around Rs. 150 per kg of green vanilla beans. The same green beans when they are processed fetch a price of around Rs. 1500 per kg. But processing technology in India for vanilla is still very primitive and many farmers are satisfied with just growing and supplying green beans. Considering the fact that cost of production is low, farmers are finding vanilla beans cultivation very attractive. In future more farmers will take up this crop and the production and export figures of vanilla will increase.

Vanilla imports are dominated by three countries- U.S.A, France and Germany. Importers in Germany and France are suppliers to other markets especially in Europe. Europe imports generally high quality beans while U.S.A accepts low quality beans also. There is an understanding between Bourbon vanilla producing countries *viz.* Madagascar, Comoro and Reunion, and importers of France and Germany in the marketing of vanilla beans.

Fenugreek

Fenugreek, a native of South Eastern Europe and West Asia, is cultivated as a leafy vegetable, condiment and as medicinal plant. The fresh tender leaves and stem are consumed as curried vegetable and the seeds are mainly used as spice for flavouring almost all dishes. It has a high medicinal value as it prevents constipation, removes indigestion, stimulates spleen and liver and is appetizing and diuretic. Botanical name of fenugreek is *Trigonella foenum-graecum L.* It is exported to Saudi Arabia, Japan, Sri Lanka, Korea and UK. The major states growing fenugreek in India are Rajasthan, Madhya Pradesh, Gujarat, Uttar Pradesh, Maharashtra and Punjab.

It is an annual herb reaching a height of about 0.9 m, leaves are light green, pinnately trifoliate, flowers – papilionaceous, fruits – legume, long, narrow, curved, tapering with a slender point and containing small deeply furrowed seeds. There are two species of the genus Trigonella, which are of economic importance *viz.*, T. foenum graecum, the common methi and T. corniculata, the Kasuri methi. These two differ in their growth habit and yield. The latter one is a slow growing type and remains in rosette condition during most of its vegetative growth period.

Climate and Soil: It has a wide adaptability and is successfully cultivated both in the tropics as well as temperate regions. It is tolerant to frost and freezing weather. It does well in places receiving moderate or low rainfall areas but not in heavy rainfall area. It can be grown on a wide variety of soils but clayey loam is relatively better. The optimum soil pH should be 6.0 to 7.0 for its better growth and development.

Varieties: Many improved cultivars are now available for cultivation. They are briefly described hereunder.

Variety	Parentage	Characters
CO1	Reselection from TG 2336	Duration of the crop is 90 days. Dual purpose quick growing, suited for intercropping, high seed protein 20-23%. Yield per ha 685kg grain, 4.51 of green.
Rajendra Kanti	Mass selection	Duration of the crop is 120 days. High yield, medium height, bushy, suited for pure as well as intercropping. Seed protein: 9.5%. yield per ha 1200 - 1400 kg grain
RMt-1	Pure line selection from Nagpur type	Duration of the crop is 145 days. High yield, moderately branched, moderately tolerant, to root rot and powder mildew. Seed protein: 21%. Yield per ha 1500 kg grain
Lam Sel.1	A selection from germplasm	Duration of the crop is 68 days. High yield, bushy plant type. Seed protein: 53%. Yield per ha 750 kg grain

Land Preparation and Sowing: Land is prepared by ploughing thrice and beds of uniform size are prepared. Broadcasting the seed in the bed and raking the surface to cover the seeds is normally followed. But, line sowing is advocated in rows at 20 to 25 cm apart which facilitates the intercultural operations.

Sowing in the plains is generally taken up in September to November while in the hills, it is grown from March. Approximately 20 to 25 kg of seed is required for one hectare and the seed takes about 6-8 days to complete its germination.

Manures and Fertilizers: Besides 15 tonnes of farm yard manure, a fertilizer dose of 25 kg Nitrogen, 25 kg Phosphorus and 50 kg Potash per ha is recommended. Half of the nitrogen dose and the entire quantity of phosphorus and potash are applied basally and the remaining half Nitrogen is applied 30 days after sowing. To obtain more successful leafy growth, nitrogen should be applied after each cutting.

Irrigation: First irrigation is given immediately after sowing and subsequent irrigation is applied at 7 to 10 days interval.

Harvesting: In about 25 to 30 days, young shoots are nipped off 4 to 5 cm above ground level and subsequent cuttings of leaves may be taken after 15 days. It is advisable to take 1 to 2 cuttings before the crop is allowed for flowering and fruiting.

When pods are dried, the plants are pulled out, dried in the sun and seeds are threshed by beating with stick or by rubbing with hands. Seeds are winnowed, cleaned and dried in the sun. They may be stored in gunny bags lined with paper.

Yield: A yield of 1200 to 1500 kg of seeds and about 800 to 100 kg of leaves may be obtained per hectare in crops grown for both the purposes.

Plant Protection: Root rot (Rhizoctonia solani) is a serious disease and can be controlled by drenching carbendazim 0.05% first at the onset of the disease and another after one month.

Clove

Clove, the dried unopened flower buds of the evergreen tree, *Syzygium aromaticum*, (Syn. *Eugenia caryophyllus*) is an important spices noted for its flavour and medicinal values. It is indigenous to Moluccas island (Indonesia) and was introduced to India around 1800 A.D. by the East Indian Company in their spice garden in Courtallam, Tamil Nadu.

The major producers of this spice today are Indonesia, Zanzibar and Madagascar. World production is estimated to be 63,700 tonnes. Indonesia alone accounts for 66 % of the world production. The important clove growing regions in India now are the Nilgiris, Tirunelveli and Kanyakumari districts of Tamil Nadu, Calicut, Kottayam, Quilon and Trivandrum districts of Kerala and South Kanara district of Karnataka.

Food processing industry uses both whole and ground form of cloves in various preparations. Clove oil is used in perfumeries, pharmaceuticals and flavouring industries. Clove oleoresin is also increasingly used in the food processing industry. In Indonesia, the major part is absorbed for making KRETEK cigarette industry. The clove is an evergreen tree often reaching a height of 7 to 15 metres. Leaves posses plenty of oil glands on the lower surface.

Climate and Soil: Clove is strictly a tropical plant and requires a warm humid climate having a temperature of 20 to 30^{0}C. Humid atmospheric condition and a well distributed annual rainfall of 150 to 250 cm are essential.

It thrives well in all situations ranging from sea level upto an altitude of 1500 metres and also in places proximal to and away from sea. Deep black loam soil with high humus content found in the forest region is best suited for clove cultivation. It grows satisfactorily on laterite soils, clay loams and rich black soils having good drainage. Sandy soil is not suitable.

Propagation: Clove is propagated through seed, which is called mother clove. The seeds become available from June to October. Fruits are allowed to ripe on the tree itself and drop down naturally. Such fruit are collected from the ground and sown directly in nursery or soaked in water overnight and the pericarp removed before sowing.

They lose their viability within one week after harvest under normal conditions and hence they must be sown immediately after collection from a tree. The second method gives quicker and higher germination. Big sized seeds generally give higher percent of germination.

Nursery Practices: Beds for sowing seeds are of 15 to 20 cm height, one metre width and a convenient length. The beds should be made of loose soils and over which a layer of sand may be spread (about 5-8cm thick). Seeds are sown at 2 to 3 cm spacing and a depth of about 2cm.

The seedbeds have to be protected from direct sunlight. The germination commences in about 10 to 15 days and may last for about 40 days. In higher elevations, germination is delayed considerably, often requires 60 days. The germinated seedlings are transplanted in polythene bags 30cm x 15 cm) containing a mixture of good soil, sand and well decomposed cowdung (in the ratio of about 3:3:1). The seedlings are ready for transplanting in the field when they are 18 to 24 months old. The nurseries are usually shaded and irrigated daily to ensure uniform stand.

Land Preparation and Planting: Eastern and northeastern hill slopes, well drained valleys and river banks are ideal for clove. The area selected for raising clove plantations is cleared of scrub growth before monsoon and pits of 60 to 75 cm 3 are dug at a spacing of 6 to 7 meters.

The pits are partially filled with topsoil. The seedlings are transplanted in the main field during the beginning of rainy seasons, in June-July, and in low lying areas, towards the end of the monsoon, in September-October. Cloves prefer partial shade.

Manuring: Clove trees are to be manured regularly and judiciously for their proper growth and flowering as given below:

Age of the plant	Cattle manure or compost (kg)	Urea	Super Phosphate	Muriate of Potash
		g/tree		
First year	15	-	-	-
Second year	20	80	220	160
Annual increase per year	5	40	110	80
Tree of 15 and above	50	600	1560	1250

The entire quantity of organic manures and half the quantity of fertilizers may be applied during May-June and the remaining quantity or fertilizers is applied in October-November in shallow trenches dug around the plant normally about 1 to 1.5m away from the tree base.

Irrigation: Irrigation is necessary in the initial stages. In places where pronounced drought is normally experienced, pot watering is recommended to save the plants in the initial two or three years. Subsoil irrigation using 20cm length mud tubes or bamboo tubes with the helpful to save the plants during acute summer. Although the trees can survive without irrigation, it is advantageous to irrigate the grown up trees for proper growth and yield.

Harvesting: Clove tree begins to yield from the seventh or eighth year after planting and full bearing stage is attained after about 15 to 20 years. The flowering season is September-October in the plains and December to February at high altitudes.

Flower buds are produced on young flush. It takes about 4 to 6 months for the buds to become ready for harvest. At this time, they are less than 2cm long. The optimum stage for picking clove buds is indicated by the change in the colour from green to slightly pinkish tinge.

The matured clove buds are carefully picked with hand. Care should be taken to pick the buds at the correct time, as otherwise the quality of the cured produce will be lost to a considerable extent. When the trees are tall and the clove bunches are beyond the reach, platform ladders are used for harvesting. Bending the branches or knocking down the bud clusters with sticks is not desirable as these practices affect the future bearing of the tree.

Curing: The harvested flower buds are separated from the clusters by hand and spread in the drying yard for drying. It takes normally 4 to 5 days for drying. The correct stage of drying is reached when the stem of the bud is dark brown and the rest of the bud lighter brown in colour. Well dried cloves will be only about one-third of the weight of the original. About 11,000 to 15,000 dried cloves make one kilogram.

Yield: Clove often experiences irregular or alternate bearing tendency. A well maintained full grown tree under favorable conditions may give 4 to 8 kg of dried buds. The average annual yields after 15th year may be taken as 2 kg per tree. Clove oil, the spice determining factor, is about 16 to 21 % in the buds. The oil contains 70 to 90 % of free eugenol and 5 to 12 % of eugenyl acetate.

Plant Protection

Diseases/Pest	Symptoms	Control Measure
Leaf rot (Cylindrocladium quinquiseptatam)	Dark patches are seen in leaves of mature trees seedlings, which often result in rotting of whole leaves or tips alone causing severe defoliation.	Spray 0.2 % carbendazim.
Stem borer	The pest bores into the main stem, causing death of the plant.	Swab the surface of the stem with 50% W.P.B.H.C.
Scale insects	Infestation is seen on leaves and tender shoots, more often in the nursery.	Spray 0.05 % monocrotophos.

Cardamom

Cardamom, popularly, known as Queen of Spices is native to the evergreen rainy forests of Western Ghats in South India. Botanical name of the cardamom is Elettaria cardamomum (L.) Maton, which belongs to the family Zingiberaceae. It is cultivated in about 1,00,000 ha mainly confined to the Southern State *viz.*, Kerala, K'taka and Tamil Nadu accounting for 60, 31 and 9% of the total area respectively. Our annual production is about 40000 metric tonnes and nearly 40% of which is exported to more than 60 countries earning a foreign exchange of nearly 60 million rupees. Cardamom is used for flavouring various preparations of food, confectionary, beverages and liquors.

Botany: Cardamom is a herbaceous perennial having underground rhizomes. The aerial pseudostem is made of leaf sheaths. Inflorescence is a long panicle with recemose clusters arising from the underground stem, but comes up above the soil. Flowers are bisexual, fragrant, fruit is a trilocular capsule. Flower initiation takes place in March-April and from initiation to full bloom, it takes nearly 30 days and from bloom to maturity, it takes about 5 to 6 months.

It is a perennial crop, propagated from the seeds or cut bits of the rhizome. It starts bearing in 3 to 5 years after planting and the economic age of the plantation is 12 to 15 years. The fruits are about 2.5 cm long, ovoid and triangular in shape brown or pink in colour when ripe. They contain 40 to 50 seeds. Average yield is 300 to 1000 kg per ha from 4th or 5th year.

Climate and Soil: The natural habitat of cardamom is the evergreen forests of Western Ghats. It is grown in the areas where the annual rainfall ranges from 1500 to 4000 mm, with a temperature range of 10 to 35^{0}C and an altitude of 600 to 1200 m above MSL. Rainfall distribution should be good and summer showers during February-April are essential for panicle initiation, otherwise it will affect the yield.

Varieties: Based on the size of the fruit, two varieties are broadly recognised *viz.*, Elettaria cardamomum var. major consisting of wild indigenous types and var. minor comprising the cultivated types *viz.*, Mysore, Malabar. Recently a number of improved cultivars have been released for cultivation.

New Cardamom Varieties

S No.	Vrietal Name	Type	Yield Potential (dry capsules/ha in kg.)	Other attributes
1.	Coog cardamom Malabar Selection-1 (CCS-1)	Malabar	408	Released by NRCS, Cardamom Research Centre, and Appangala an o.p. Seedling of clone 37. Dry recovery 22%, Essential oil content: 8.7%.
2.	ICRI-1	Malabar	325-rainfed 650-irrigated	Released by Indian Cardamom Research Institute, Myladumpara. Dry recovery: 22.9%, Essential oil: 8.3%.
3.	ICRI-2	Mysore	375-rainfed 760-irrigated	Released by ICRI, Myladumparai, adaptable for Vandenumedu and Anamalai hills. Dry recovery: 22.5%, Essential oil: 9%.
4.	Mudigree-1	Malabar	250-300	Released by RRS, Mudigree suitable for Malned region of Karnataka State. Dry recovery: 20%, Essential oil: 8%.
5.	PV-1	Malabar	500	Released by Cardamom Research Station, Pampadumpara. Dry recovery: 20%, Essential oil: 6.8%.
6.	SKP-14	Malabar	430 rainfed 590-irrigated	Released by ICRI, Regional Station, Saklespur, and Karnataka.

Nursary Management: Cardamom is propagated mainly through seeds and also through suckers each consisting of atleast one old and a young aerial shoot. The suckers are commonly used for gap filling but suckers may not be available in larger numbers. Therefore, a rapid clonal multiplication technique evolved by N.R.C.S., Cardamom Reserch Centre, Appangala, is proved to be quick, reliable and economic for production of large number of quality planting materials. The site selected for this method should have a gentle slope and must be nearer to the water source.

Trenches of 45cm width, 45cm depth and of any convenient length may be taken across the slope or along the contour at 1.8 m apart. The top 20cm depth soil is excuvated separately and heaped on the upper side of the trench. The lower 25 cm depth soil is excavated and heaped on lower side of the trenches all along the line. The top soil is mixed with equal protions of humus rich jungle soil, sand and cattle manure and filled back by leaving a depression of 5cm at the top to facilitate mulching for retention of soil mixture. Suckers, each consisting of one grown up tiller and a growing young shoot, are placed at a spacing of 0.6 m in the trenches during March-October.

Regular cultural operations are to be followed including a high fertilizer dose of 100:50:200 kg NPK/ha in 6 split doses at 60 days interval along with neem cake at 250 g/plant.

Irrigation should be provided atleast twice a week. Over head pandal at a height of 3.6m covered with coir mat or leafy twigs of any shade tree may be provided during the non-rainy season or pandal at a height of 3.6m covered with coirmat. Within a period of 12 months, a plant would produce atleast 32 to 42 suckers which may yield at least 16 to 21 planting units. *i.e.* about 1.5 lakhs planting units per ha. of clonal nursery within 12 months of planting.

Seedlings are normally raised in primary and secondary nurseries. The nursery site should be selected on gentle slopy lands, having an easy access to a water source. Raised beds are prepared after digging the lands to a depth of 30-45 cm. The beds of 1m width and convenient length raised to a height of about 30cm are prepared. A fine layer of humus-rich forest soil is spread over the beds. Seeds are to be collected from well ripe capsules. Immediately after harvesting, the husk is removed and the seeds are washed repeatedly in water for removing the mucilaginous coating. After draining the water, the seeds are to be mixed with wood ash and dried in shade for a day. In order to ensure uniform and early germination, seeds should be sown immediately after extraction. If the sowing is delayed, pre-sowing treatment of seeds with 25% Nitric acid for 10 minutes is advisable to get a quick and higher germination. One kg of seed capsules may produce 5000 seedlings.

Sowing may be taken up during November-January and is done in rows. Deep sowing of seeds has to be avoided for better and quick germination. Seedbeds are to be dusted with BHC 10%. Beds are mulched to a thickness of 2 cm with paddy straw or any locally available material and are watered regularly. The germination commences in about 30 days and may continue for a month or two. After germination, the mulch is to be removed. As overhead pandal with a height of 2 m is quite desirable. Materials like coir mat, plaited coconut leaves or tree twigs, which do not shed their leaves easily, may be used but the coir mat is preferred as it allows uniform filtered sunlight.

The excess seedlings are to be thinned out after 75-80 days of sowing. The thinned out seedlings may be used for gap filling within the nursery bed or for raising secondary nursery. When the seedlings attain 5-6 leaf stage, light earthing up is to be done. This would encourage better tillering and proper growth of seedlings. Generally in Kerala and Tamil Nadu, the seedlings are transplanted to the secondary nursery when they attain four to six leaf stage. The beds are prepared in the same manner as that of a primary nursery. Seedlings are transplanted in March-May at a spacing 20x20 cm and mulched immediately. Beds are to be covered with an over head pandal and should be watered regularly. Recently, instead of secondary nursery beds, the seedlings are also raised in polybags containing rich forest soil.

Manuring at the rate of 90g Nitrogen, 60g Phosphorus and 120g Potash per bed of 5x1m size, in three equal split doses at an interval of 45 days is recommended to produce healthier seedlings. The first dose of fertilizer may be applied 30 days after transplanting in the secondary nursery. In K'taka, 10 months old seedlings are used for planting in the main yield, while in Kerala and Tamil Nadu, 18 months old seedings are commonly used.

Preparation of Land: All under growth should be cleared and excess shade trees or branches should be thinned out to have an even overhead canopy. Pits of 45 x 45 x 30-cm size are dug in April-May and filled with a mixture of topsoil and compost or well-decomposed farmyard manure. In slopy land, contour terraces may be made and pots may be taken along the contour. The spacing adopted in K'taka for the Malabar type is 2 x 2 m between plants and rows. In Kerala region 2-3 on either side is adopted. Staggered trenches may be taken across the slope to conserve run off rainwater. The soil collected in trenches may be utilised for earthing up during the post-monsoon period.

Planting: The planting is carried out during the rainy season commencing from June. Under Eastern Ghat hills, July planting is adopted. Seedlings are to be planted upto the collar region for better growth. Cloudy days with light drizzle are ideal for planting.

Mulching: It is an important cultural practice in cardamom. Fallen leaves of the shade trees are utilised for mulching.

Sufficient mulch should be applied during November-December to reduce the ill effects of drought, which prevails for nearly 4 to 5 months during summer. Exposing the panicle over the mulch is beneficial for pollination by the bees.

Weeding: Depending upon the intensity of weeds, 2 to 3 rounds of weeding are necessary in a year. The first round of weeding is to be carried in May-June, the second in August-September and the third in December-January. In slopy land slashing of weeds is alone to be carried out otherwise it encourages to more soil erosion. Weedicides like paraquat @625 ml in 500 litres of water may be sprayed in the interspaces between rows leaving 60 cm around the plant base.

Trashing: Trashing consists of removing old and drying shoots of the plant once in a year with the onset of monsoon under rainfed conditions and 2-3 times in high density plantations provided with irrigation facilities.

Shade Regulation: Cardamom being a pseophyte, is very sensitive to moisture stress. Shade helps to regulate soil moisture as well as temperature and provides congenial microclimate for cardamom. Excess shade is also quite detrimental and shade has to be regulated so as to provide 50-60% filtered sunlight. Cardamom plants can tolerate less shade in areas where well-distributed and adequate rainfall is received. In South India, many trees are available in the natural habitat to provide shade but an ideal shade tree should have a wider canopy, minimum side branching and it should not shed the leaves during flowering phase of Cardamom so as not to affect pollination. Some of the common shade trees in Cardamom estates are Palangi (Artocarpus frazinifolius), Jack, Red cedar (Cedrella toona), Karimaram (Diospyros ebenum) and Elangi (Mimusops elangi). The temporary shade trees like Erythrina lithosperma and E. indica are most unsuitable as they compete for nutients and soil moisture.

In order to provide adequate light during monsoon, shade regulation may be taken up before the onset of monsoon. A two-tier canopy with a height of not more than 3 m between the lower and higher canopy may be maintained. Areas exposed to western side should have adequate shade.

Earthing Up: After the monsoon is over, a thin layer of fresh fertile soil, rich in organic matter may be earth up at the base of the clump, covering up to the collar region by scraping between the rows or collecting soil from staggered trenches/ check pits. This encourages new growth.

Irrigation: In order to overcome the day spell during summer, it is necessary to irrigate the crop to get maximum production as it helps in initiation of panicles, flowering and fruitset. Depending on the moisture holding capacity of soil and topography of the estates, they may be irrigated at an interval of 10 to 15 days till the onset of monsoon. Sprinkler irrigation and or drip irrigation at the rate of 4 litres per clump per day during dry months increases the yield.

Manuring: A fertilizer does of 75 kg Nitrogen, 75kg Phosphorus and 150 kg Potash per ha is recommended under irrigated condition for high yielding plantations yielding 100kg/ ha and above and a dose of 30:60:30 kg/ha is recommended for gardens under rainfed condition. Besides, organic manures like compost or cattle manure may be given @ 5kg per clump. Neem oil cake may also be applied @ one kg per clump. Foilar application of urea at 3%, single super phosphate at 1%, and muriate of potash at 2% is also beneficial to improve the yield. In places were zinc deficiency is noticed, spraying of zinc 500 ppm twice in a year, April-May and September-October is recommended.

Fertiliser is applied in two split doses. The first application during May will help in the production of suchers and development of capsules and the second application during late September will help the initiation of panicles and sucker. Only half the dose of fertilizer is to be applied during the first year and full does is given from second year onwards. Being a surface feeder, deep placement of fertiliser is not advocated. Application of fertiliser is done at a radius of 30cm and covered with a thin layer of soil.

Harvesting and Processing: Cardamom plants normally start bearing two years after planting. In most of the areas the peak period of harvest is during October-November. Picking is carried out at an interval of 15-25 days. Ripe capsules are harvested in order to get maximum green colour during curing.

After harvest, capsules are dried either in fuel kiln or electrical drier or in the sun. It has been found that soaking the freshly harvested green cardamom capsules in 2% washing soda solution for 10 minutes prior to drying helps to retain the green colour during drying. When drier is used, it should be dried at 45 to 50^{0}C for 14 to 18 hours, while for kiln, over night drying at 50 to 60^{0}C is required. The capsules kept for drying are spread thinly and stirred frequently to ensure uniform drying. The dried capsules are rubbed with hands or coir mat or wire mesh and winnowed to remove any foreign matter. They are then sorted out according to size and colour, during storage. These bags are then kept in wooden chambers.

Plant Protection: Important pests and diseases affecting cardamom are given below with their typical damages/symptoms and control measures.

Pest/Disease	Damage/symptoms	Control measures
Pests Thrips (Sciothrips acrdamomi)	Adults and nymphs such sap and cause damage to leaves, shoots, inflorescence, and thrips affected capsules fetch lower price.	Regulate shade in thickly shaded area, spray monocrotophos 0.025% during March to September
Shoot, panicle, capsule/borer (Conogethes punctiferalis)	Larvae bore the unopened leaf buds, panicles causing drying of feed on young seeds causing the capsules empty, incidence occur throughout year, but more pronounced during March-April, May-June and September-October.	Spraying monocrotophos or fenthion 0.075% at early stage of infection.
Aphids (Pentalonia nigronervosa)	Nymphs and adults suck the sap and act as vector of the mosaic or 'Katte' Virus.	Spray 0.05% dimethoate.
Parasitic nematodes (Meloidogyne incognita)	Occurs in nursery and main field. Poor germination and establishment in the nurseries, stunted and poor growth of plants, shedding of immature capsules in the main field, heavy galling and abnormal branching of roots.	Treat the plantsin the nursery with carbofuran 3 g @ 5 kg a.i./ha or in the main field with carbofuran 5-g a.i./clump and apply 0.5kg of neem cake per clump twice a year.
Diseases Katte diseases (Viral disease)	Spindle shaped, slender chlorotic flecks appear on youngest leaves, later there develop into pale green discontinuous stripes as leaves mature, mosaic symptoms are marked, infected clumps are stunted, smaller in size, with slender tillers and shorter panicles.	Use healthy seedlings, rougue the infected plants.
Azhukal capsule rot (Phytophthora meadii) (P.nicotianae var. nicotianae)	Occurs in heavy rain zones, affected capsules turn brownish black in colour, often rotting extends to tillers and rhizomes also.	Do trashing, remove infected and dead plants etc. During premonsoon months. Spray 1-% Bordeux mixture during May and repeat again in August.
Damping or rhizome rot (Pythium vexans and (Rhizoctonia solani)	Excessive soil moisture and poor drainage favour this disease in the nursery, infected seedlings collapse at collar region and die in patches, entire clump dies in grown up plants.	Pre treat the nursery with 1:50 formaldehyde, drench the soil after germination with 0.2% copper oxychloride.

Introduced Species

Sweet clover (*Melilotus sp.*), introduced and naturalized to the U.S. from Eurasia as a forage and cover crop.

An introduced species (also known as naturalized species or exotic species) is an organism that is not indigenous to a given place or area and instead has been accidentally or deliberately transported to this new location by human activity. Introduced species can often be damaging to the ecosystem it is introduced to. A list of introduced species is given in a separate article.

Terminology

The terminology associated with introduced species is presently in flux for a variety of reasons. Other terms that are used sometimes interchangeably (having the same or similar meanings) with *introduced* are *acclimatized, adventive, alien, bioinvasive, exotic, escaped, feral, foreign, invasive, non-native, naturalized, immigrant, non-indigenous*, and *xenobiotic*. Nonetheless, distinctions can and should be made between some of these terms.

In the broadest and most widely used sense, an introduced species is synonymous with *non-native* and therefore applies as well to most garden and farm organisms; these adequately fit the basic definition given above. However, some sources add to that basic definition: "...and are now reproducing in the wild", which removes from consideration as *introduced* all of those species raised or grown in gardens or farms that do not survive without tending by people. With respect to plants, these latter are in this case defined as either *ornamental* or *cultivated* plants.

The following definition from the United States Environmental Protection Agency, although perhaps lacking ecological sophistication, is more typical: *introduced species* are.."[s]pecies that have become able to survive and reproduce outside the habitats where they evolved or spread naturally".

There is valid disagreement as to whether the term invasive species is exactly synonymous with *introduced species*. A species that is *invasive* is one that has been introduced and become a pest in its new location, spreading (invading) by natural means. The term is used to imply both a sense of urgency and actual or potential harm. For example, U.S. Executive Order 13112 (1999) defines "invasive species" as "an alien species whose introduction does or is likely to cause economic or environmental harm or harm to human health".

Although some argue that "invasive" is a loaded word and harm is difficult to define, the fact of the matter is that organisms have and continue to be introduced to areas where they are not native, sometimes with, usually without, much regard to the harm that could result. Ecological purists would argue that all non-natives capable of becoming established in the wild are harmful where they are introduced.

From a regulatory perspective, it is neither desirable nor practical to simply list as undesirable or outright ban all non-native species (although the State of Hawaii has adopted an approach that comes close to this). Regulations require a definitional distinction between non-natives that are deemed especially onerous and all others. Introduced *pest species* that are *officially listed* as invasive, best fit the definition of an *invasive species*.

Table of terms related to "Introduced Species"

<table>
<tr><td colspan="4">Table of terms related to "Introduced Species"</td></tr>
<tr><td>NATIVE</td><td colspan="3">NON-NATIVE
INTRODUCED (broad definition)</td></tr>
<tr><td rowspan="2">INDIGENOUS or ENDEMIC</td><td rowspan="2">CULTIVATED and LIVESTOCK</td><td colspan="2">Established in the wild INTRODUCED (narrow definition)</td></tr>
<tr><td>INVASIVE (pest)</td><td>All others not listed*</td></tr>
</table>

*Not listed in any "official" source as a pest species

The Nature of Introductions

By definition, a species is considered "introduced" when its transport into an area outside of its native range is human mediated. Introductions by humans can be described as either intentional or accidental. Intentional introductions have been motivated by individuals or groups who believe that the newly introduced species will be in some way beneficial to humans in its new location. Unintentional or accidental introductions are most often a byproduct of human movements, and are thus unbound to human motivations. Subsequent range expansion of introduced species may or may not involve human activity.

Intentional Introductions

Species that humans intentionally transport to new regions can subsequently become successfully established in two ways. In the first case, organisms are purposely released for establishment in the wild. It is sometimes difficult to predict whether a species will become established upon release, and if not initially successful, humans have made repeated introductions to improve the probability that the species will survive and eventually reproduce in the wild. In these cases it is clear that the introduction is directly facilitated by human desires.

In the second case, species intentionally transported into a new region may escape from captive or cultivated populations and subsequently establish independent breeding populations. Escaped organisms are included in this category because their initial transport to a new region is human motivated.

Perhaps the most common motivation for introducing a species into a new place is that of economic gain. Examples of species introduced for the purposes of benefiting agriculture, aquaculture or other economic activities are widespread. Eurasian carp was first introduced to the United States as a potential food source. The apple snail was released in Southeast Asia with the intent that it be used as a protein source, and subsequently to places like Hawaii to establish a food industry. In Alaska, foxes were introduced to many islands to create new populations for the fur trade.

The timber industry promoted the introduction of Monterey Pine (*Pinus radiata*) from California to Australia and New Zealand as a commercial timber crop. These examples represent only a small subsample of species that have been moved by humans for economic interests.

Introductions have also been important in supporting recreation activities or otherwise increasing human enjoyment. Numerous fish and game animals have been introduced for the purposes of sport fishing and hunting. The introduced amphibian (*Ambystoma tigrinum*) that threatens the endemic California salamander (*Ambystoma californiense*) was introduced to California as a source of bait for fishermen. Pet animals have also been frequently transported into new areas by humans, and their escapes have resulted in several successful introductions, such as those of feral cats and parrots.

Many plants have been introduced with the intent of aesthetically improving public recreation areas or private properties. The introduced Norway Maple for example occupies a prominent status in many of Canada's parks. The transport of ornamental plants for landscaping use has and continues to be a source of many introductions. Some of these species have escaped horticultural control and become invasive. Notable examples include water hyacinth, salt cedar, and purple loosestrife.

In other cases, species have been translocated for reasons of "cultural nostalgia," which refers to instances in which humans who have migrated to new regions have intentionally brought with them familiar organisms. Famous examples include the introduction of starlings to North America by Englishman Eugene Scheiffer, a lover of the works of Shakespeare, who wanted to introduce all of the birds mentioned in Shakespeare's plays into the United States. He deliberately released eighty starlings into Central Park in New York City in 1890, and another forty in 1891. Yet another prominent example is the introduction of the European rabbit to Australia by one Thomas Austin, a British landowner who had the rabbits released on his estate in Victoria because he missed hunting them. A more recent example is the introduction of the wall lizard to North America by a Cincinnati boy, George Rau, in the 1950s.

Intentional introductions have also been undertaken with the aim of ameliorating environmental problems. A number of fast spreading plants such as Garlic Mustard and kudzu have been introduced as a means of erosion control. Other species have been introduced as biological control agents to control invasive species and involves the purposeful introduction of a natural enemy of the target species with the intention of reducing its numbers or controlling its spread.

A special case of introduction is the reintroduction of a species that has become locally endangered or extinct, done in the interests of conservation. Examples of successful reintroductions include wolves to Yellowstone National Park in the U.S., and the Red kite to parts of England and Scotland. Introductions or translocations of species have also been proposed in the interest of genetic conservation, which advocates the introduction of new individuals into genetically depauperate populations of endangered or threatened species.

The above examples highlight the intent of humans to introduce species as a means of incurring some benefit. While these benefits have in some cases been realized, introductions have also resulted in unforeseen costs, particularly when introduced species take on characteristics of invasive species.

Accidental Introductions

Unintentional introductions occur when species are transported by human vectors. For example, three species of rat (the Black, Norway and Polynesian) have spread to most of the world as hitchhikers on ships. There are also numerous examples of marine organisms being transported in ballast water, one being the zebra mussel. Over 200 species have been introduced to the San Francisco Bay in this manner making it the most heavily invaded estuary in the world. Increasing rates of human travel are providing accelerating opportunities for species to be accidentally transported into areas in which they are not considered native.

Introduced Plants and Algae

Many non-native plants have been introduced into new territories, initially as either ornamental plants or for erosion control, stock feed, or forestry. Whether an exotic will become invasive is seldom understood in the beginning, and many non-native ornamentals languish in the trade for years before suddenly naturalizing and becoming invasive. Peaches, for example, originated in Persia, and have been carried to much of the populated world. Tomatoes are native to the Andes. Squash (pumpkins), maize, and tobacco are native to the Americas, but were introduced to the Old World. Many introduced species require continued human intervention to survive in the new environment. Others may become feral, but do not seriously compete with natives. They simply increase the Bio-diversity of the area.

Dandelions are also Introduced Species to North America

A very troublesome marine species in southern Europe is the seaweed *Caulerpa taxifolia*. *Caulerpa* was first observed in the Mediterranean Sea in 1984, off the coast of Monaco. By 1997, it had covered some 50 km^2. It has a strong potential to overgrow natural biotopes, and represents a major risk for sublittoral ecosystems. The origin of the alga in the Mediterranean was thought to be either as a migration through the Suez Canal from the Red Sea, or as an accidental introduction from an aquarium.

Japanese knotweed grows profusely in many nations. Human beings introduced it into many places in the 19th century. It is a source of resveratrol, a dietary supplement.

Introduced Animals

One of the most egregious examples of introducing an exotic animal was perpetrated by one Eugene Scheiffer, a lover of the works of Shakespeare, who wanted to introduce all of the birds mentioned in Shakespeare's plays into the United States. He deliberately released eighty starlings into Central Park in New York City in 1890, and another forty in 1891. The starling had been introduced previously into Ohio and had failed to survive.

Other examples of introduced animals include the gypsy moth in eastern North America, the zebra mussel and alewife in the Great Lakes, the Canada Goose and Gray Squirrel in Europe, the Muskrat in Europe and Asia, and the Common Brushtail and Common Ringtail Possums in New Zealand.

Invasive Exotic Diseases

History is rife with the spread of exotic diseases, such as the introduction of smallpox into the Americas, where it obliterated entire Native American civilizations before they were ever even seen by Europeans.

Problematic exotic disease introductions in the past century or so include the chestnut blight which has virtually extinguished the American chestnut, and Dutch elm disease, which has severely damaged the American elm.

The most Commonly Introduced Species

Some species, such as the Brown Rat, House Sparrow, Ring-necked Pheasant and European Starling, have been introduced very widely. In addition there are some agricultural and pet species that frequently become feral; these include rabbits, dogs, goats, fish, pigs and cats. It is also hard to believe that something as common as a dandelion is an alien species that was imported here at some point in time.

Introduced Species on Islands

Perhaps the best place to study problems associated with introduced species is on islands. Depending upon the isolation (how far an island is located from continental biotas), native island biological communities may be poorly adapted to the threat posed by exotic introductions. Often this can mean that no natural predator of an introduced species is present, and the non-native spreads uncontrollably into open or occupied niche.

An additional problem is that birds native to small islands may have become flightless due to the absence of predators prior to introductions, and cannot readily escape danger. The tendency of rails in particular to evolve flightless forms on islands has led to the disproportionate number of extinctions in that family.

The field of island restoration has developed as a field of conservation biology and ecological restoration, a great deal of which deals with the eradication of introduced species.

New Zealand

In New Zealand the largest commercial crop is *Pinus radiata*, the Monterey Pine from California, which grows better in New Zealand than in California.

However, the pine forests are also occupied by deer from North America and Europe and by possums from Australia. All are exotic species and all have thrived in the New Zealand environment. The pines are seen as beneficial while the deer and possums are regarded as serious pests.

Common gorse, originally a hedge plant in Scotland, was introduced to New Zealand for the same purpose. Like the radiata pine, it has shown a favour to its new climate and is regarded as a noxious plant which threatens to obliterate native plants in much of the country and is hence routinely eradicated, though it can also provide a nursery environment for native plants to re-establish themselves.

Rabbits, introduced as a food source by sailors in the 1800s, have become a severe nuisance to farmers, notably in South Island. The myxomatosis virus was illegally imported and illegally released but it had little lasting effect upon the rabbit population other than to make it more resistant to the virus.

Rats, brought either by the second population group to arrive in New Zealand (the Maori) or by Europeans have had a devastating effect upon native birdlife, particularly as many New Zealand birds are flightless.

Feral cats and dogs which were originally brought as pets are also known to kill large numbers of birds. A recent (2006) study in South Island has shown that even domestic cats with a ready supply of food from their owners may kill hundreds of birds in a year, including natives.

Sparrows, which were brought to control insects upon the introduced grain crops, have displaced native birds as have rainbow lorikeets and cockatoos (both from Australia) which fly free around areas west of Auckland City such as the Waitakere Ranges.

In much of the New Zealand the Australian black swan has effectively eliminated the existence of the previously introduced mute swan.

Two notable varieties of spiders have also been introduced: the white tail spider and the black widow spider. Both may have arrived inside shipments of fruit. Prior to this the only spider (and the only poisonous animal) dangerous to humans was the native katapo which is very similar to the black widow and which is known to successfully interbreed with the more aggressive North American variety.

Invasive Species

The term invasive species refers to a subset of introduced species or non-indigenous species that are rapidly expanding outside of their native range. Invasive species can alter ecological relationships among native species and can affect ecosystem function and human health. A species is regarded as *invasive* if it: (1) has been introduced by human action to a location where it did not previously occur naturally, (2) becomes capable of establishing a breeding population in the new location without further intervention by humans, and (3) spreads widely throughout the new location. In simple terms, an invader has to (1) arrive, (2) survive, and (3) thrive.

The species must find a vector that will bring it to a new environment. This new habitat must be a close enough match to its native range that it is able to survive and reproduce here without human cultivation (Williams and Meffe, 1998). To actually become invasive, the introduced species has to be able to out-compete native species, to reproduce effectively enough to start spreading geographically through its new habitat, and to negatively impact the ecosystems in its introduced range.

Conditions that Lead to Invasion

Scientific literature proposes several mechanisms to explain the success of invasive species. These mechanisms generally fall into two different categories: species-based mechanisms and ecosystem-based mechanisms. More likely, it is a combination of several mechanisms that cause an invasive situation to occur.

Species-based Mechanisms

Species-based characteristics focus on plant competition. While all plants are able to compete in some manner to survive and oppose, invasive species appear to have specific traits or combinations of specific traits that make them especially good competitors. In some cases it can be as simple as having the ability to grow and reproduce more rapidly than native species. Other situations are more complex, such as allelopathy, which is a common mechanism whereby the invader directly or indirectly prevents other plants from growing nearby.

Life History

The life history of an organism describes the different stages of life an organism will go through during its lifetime. Such traits are tempting to study because life history is a quantifiable trait that could lead to very predictive models.

Several traits have been singled out by researchers as predictors of invasive ability in plants. For example, the ability to reproduce both asexually (vegetatively) as well as sexually, rapid growth, early sexual maturity, high reproductive output, the ability to disperse young widely, tolerance of a broad range of environmental conditions, and high phenotypic plasticity are all abilities that might aid an invasive plant in establishing and proliferating in a new environment.

The majority of studies agree in a general sense on which kind of traits mark an invasive species, but there are differences in how invasive each trait can make a species. One study found that of a list of invasive and noninvasive species, 86% of the invasive species could be identified from the traits alone.

Another study found that invasive species tended to only have a small subset of the invasive traits, and that many of these invasive traits were found in non-invasive species as well. This is one of the great difficulties in invasive species research: while many generalities can be made about invasive species, there are always exceptions to these observations (Kolar and Lodge 2001, Thebaud *et al.* 1996, Reichard and Hamilton 1997).

Superior Competition

A common trait of invasive species is great competitive ability, which can be stronger against plants in a new habitat than plants in their native habitat. There can be huge differences between how an invasive species interacts with its native ecosystem, and with the ecosystem it is invading. Often, the invading species has a better chance at acquiring resources, such as light, water, space, or nutrients. Ecosystems where all available resources are being used to their full capacity by native plants can be modeled as zero-sum systems, where any gain for the invader is a loss for the native. However, such unilateral competitive superiority (and instant, equivalent extinction of native plants with increased populations of the invader) is not the rule (Stolgren 2003, Sax *et al.* 2002). Invasive plants can coexist with native plants for an extended time, and only gradually does the superior competitive ability of an invasive species become apparent, as its population grows larger and denser, and slowly increases the risk of extinction to other species.

An invasive species might be able to use resources previously unavailable to native plants, such as very deep water sources accessed by a long taproot, or an ability to live on previously uninhabited soil types. For example, barb goatgrass (*Aegilops triuncialis*), can be found in its introduced range in California on serpentine soils, which have low water-holding capacity, low nutrients, high Mg/Ca ratio, and possible heavy metal toxicity. Few plants have adapted to grow on them, which may explain why they have had so few plant invasions. However, since goatgrass can take advantage of these soils, it can also use the light, water, and space resources that other plants are restricted from using (Huenneke *et al.* 1990). By using these resources, goat grass can become so dense as to exclude other species and form monospecific swards.

There are other reasons that an invader might be a superior competitor. For example, an invasive plant may be inedible to local herbivores, allowing it to flourish unmolested where the native species are constantly held in check. The herbivores would find themselves in increasing competition with each other over fewer native plants, while the invader is taking the place of the native species. As the invader comes to dominate its new habitat, the local food webs are changed from the bottom up, since their foundation of native plants has been altered. (Petren and Case 1996, Gray 1986).

Facilitation

Facilitation is the mechanism by which some species can alter their environment through chemicals or manipulation of abiotic factors, usually to make it more favorable to their growth or reproduction. Sometimes, neighbouring species may benefit by another's facilitation, but often the facilitation actually benefits the target species to the detriment of its neighbours. One such facilitative mechanism is allelopathy, also known as chemical competition. In allelopathy, a plant will secrete chemicals which make the surrounding soil uninhabitable, or at least inhibitory, to other plant species.

One example of this is the knapweed (*Centaurea diffusa*). This Eastern European weed has spread its way through the western United States. Experiments show that 8-hydroxyquinoline, a chemical produced at the root of *C. diffusa*, has a negative effect only on plants that have not co-evolved with *C. diffusa*.

Such co-evolved native plants have also evolved defences, and *C. diffusa* does not appear in its native habitat to be an overwhelmingly successful competitor. This result shows how difficult it can be to predict whether a species will be invasive just from looking at its behaviour in its native habitat, and demonstrates the potential for novel weapons to aid in invasiveness (Vivanco *et al.* 2004, Hierro and Callaway 2003).

Changes in fire regimes are another form of facilitation. *Bromus tectorum*, originally from Eurasia, is highly fire-adapted. It not only spreads rapidly after burning, but actually increases the frequency and intensity (heat) of fires, by providing large amounts of dry detritus during the height of the fire season in western North America. In areas where it is widespread, it has altered the local fire regime so much that native plants cannot survive the frequent fires, allowing *B. tectorum* to further extend and maintain dominance in its introduced range (Brooks *et al.* 2004).

Facilitation also occurs when one species physically modifies a habitat and that modification is advantageous to other species. For example, zebra mussels increase habitat complexity on lake floors providing nooks and crannies in which invertebrates live. This increase in complexity, together with the nutrition provided by the waste products of mussel filter-feeding increases the density and diversity of benthic invertebrate communities (Silver Botts *et al.* 1996).

Ecosystem based Mechanisms

Unused Resources

When examining an ecosystem, it is important to look at two things: the amount of resources available, and how much of those resources are being used. In a stable ecosystem, all of the available resources are used up, and all of the species there have enough to continue to survive. Yet what happens if spare resources happen to appear in an ecosystem? For example, lets say a forest fire has cleared an area, leaving only hardy brush and established trees. The resources used by the cleared species (*e.g.* grasses) are no longer being utilized, and may be taken up by other organisms. While these resources are most likely to be used by the nearby species, there is an opportunity for exotic species to establish themselves. The data shows that nitrogen and phosphorus are often the limiting factors for a situation such as this (Davis et al 2000).

Unfilled Niches

Every species has a role to play in its native ecosystem; some are general while others are highly specialized. These roles are known as niches. This mechanism describes a situation where the invaded ecosystem in question has unfilled niches, which are promptly filled by an invader.

Ecosystem Instability

Imagine a situation involving a group of people who have mastered a game, and are able to beat anyone outside of their group. They know the rules inside and out, and are constantly perfecting their strategies to win. Few, if any, have a chance against this group.

Then suddenly the rules change, and this group is at a loss, since their strategies no longer work the way they used to, and other players with vastly different strategies are able to come in, and win where they could not before. This is how the familiar ecosystem mechanism works.

This mechanism describes a situation where the ecosystem in question has suffered a disturbance of some sort, which changes the fundamental nature of the ecosystem (Byers 2002). Examples of this type of disturbance can range from the loss of an important predator to the eutrophication of an aquatic environment. In situations such as these, the specialized evolution of the native plant species becomes wasted, and non-native species can gain a foothold.

Ecology of Invasive Species

Ecological Circumstances of Invasive Species

Although an invasive species is often defined as an introduced species that has spread widely and causes harm, some species native to a particular area can, under the influence of natural events such as longterm rainfall changes or human modifications to the habitat, increase in numbers and become invasive. The Pied Currawong of southeast Australia is an example; as a result of human changes to the landscape, Pied Currawongs increased greatly in range during the 20th century and have caused substantial declines in the populations of the smaller birds whose nestlings they prey on.

All species on Earth go through periods of increasing and decreasing population numbers, in many cases accompanied by expansion and contraction of range. Human "alterations" on the landscape are especially significant. Anthropogenic alteration of an environment may enable the expansion of a species into a geographical area where it had not been seen before and thus that species could be described as invasive because the range expansion results in the species occurring where it was not before native.

In essence, one must define "native" with care, as it refers to some natural geographic range of a species, and is not coincident with human political boundaries. Whether noticed increases in population numbers is sufficient reason to regard a native species as "invasive" requires a broad definition of the term—but it seems reasonable to consider that some native species in disrupted ecosystems can spread widely and cause harm and in that sense become invasive.

Traits of Invasive Species

Many features have been attributed to invasive species and invaded ecosystems, but none are universal and invasive species tend to have a suite of traits rather than all of them. Common invasive species traits include fast growth, rapid reproduction, high dispersal ability, phenotypic plasticity (the ability to alter one's growth form to suit current conditions), tolerance of a wide range of environmental conditions, ability to live off of a wide range of food types, asexual reproduction, and association with humans (Williams and Meffee 1998). The single best predictor of invasiveness, however, is whether or not the species has been invasive elsewhere (Ewel *et al.* 1999).

A study done by Marcel Rejmanek and David Richardson (1996) on invasive Pine species in South Africa found three main traits that successfully separated invasive from non-invasive species. These are: small seed sizes, early reproduction, and short periods between large seed sets.

Before an invasive species can even put some of the aforementioned "invasive traits" to work and thrive, it must survive at a very low population density when it may be difficult to find mates or cross-fertilize or when random changes in the environment could easily wipe out the entire population (Tilman 2004).

This is why successful invasive species are often associated with humans. Our repeated patterns of movement, such as ships sailing to and from ports, or cars driving up and down highways, allow for species to have multiple opportunities of establishing (also known as a high "propagule pressure") (*e.g.*, Verling *et al.* 2005).

Traits of Invaded Ecosystems

In 1958, Charles S. Elton argued that ecosystems with higher species diversity were less subject to invasive species because of fewer available niches. Since then, other ecologists have pointed to highly diverse, but heavily invaded ecosystems and have argued that ecosystems with high species diversity seem to be more susceptible to invasion (Stohlgren *et al.* 1999).

In the end, this debate seems largely to hinge on the spatial scale at which invasion studies are performed, and the issue of how diversity affects community susceptibility to invasion remains unresolved. Small-scale studies tend to show a negative relationship between diversity and invasion, while large-scale studies tend to show a positive relationship. The latter result may be an artifact of invasive or non-native species capitalizing on increased resource availability and weaker overall species interactions that are more common when larger samples are considered (Levine 2000, Byers and Noonberg 2003)

Invasion is more likely if an ecosystem is similar to the one that the potential invader evolved (Williams and Meffee 1998). Island ecosystems may be prone to invasion because their species are "naive" and have faced few strong competitors and predators throughout their existence, or because their distance from colonizing species populations makes them more likely to have "open" niches (Stachowicz and Tilman 2005).

An example of this phenomenon is the decimation of the native bird populations on Guam by the invasive brown tree snake (Fritts and Leasman-Tanner 2001).

Alternately, invaded ecosystems may lack the natural competitors and predators that keep introduced species in check in their native ecosystems, a point that is also seen in the Guam example. Lastly, invaded ecosystems have often experienced disturbance, usually human-induced (Williams and Meffee 1998).

This disturbance may give invasive species, which are not otherwise co-evolved with the ecosystem, a chance to establish themselves with less competition from more adapted species (Tilman 2004).

Vectors

Non-native species have many *vectors*, including many natural ones, but most of the species that we consider "invasive" are associated with human activity. Natural range extensions are common in many species, but the rate and magnitude of human-mediated extensions in these species tend to be much larger than natural extensions, and the distances that species can travel to colonize are also often much greater with human agency (Cassey *et al.* 2005).

One of the earliest human influenced introductions involves prehistoric humans introducing the Pacific rat (*Rattus exulans*) to Polynesia (Matisoo-Smith *et al.* 1998). Today, non-native species come from horticultural plants either in the form of the plants themselves or animals and seeds carried with them, from animals and plants released through the pet trade. Invasives also come from organisms stowed away on every type of transport vehicle imaginable, to name a few unintentional vectors. For example, ballast water taken up at sea and released in port is a major source of exotic marine life.

Species have also been introduced intentionally. For example, to feel more "at home", American colonists formed "Acclimation Societies" that repeatedly released birds that were native to Europe until they finally established along the east coast of North America. As a result, several native bird species were pushed to extinction.

Economics play a major role in exotic species introduction. The scarcity and demand for the valuable Chinese mitten crab is one explanation for the possible intentional release of the species in foreign waters.

Impacts of Invasive Species

Ecological Impacts

Biological species invasions can negatively impact ecological systems in a multitude of ways. Worldwide an estimated 80% of endangered species could suffer losses due to competition with or predation by invasive species (Pimentel *et al.* 2005). As highly adaptable and generalized species are introduced to environments already impacted by human activities, native species are put at a distinct disadvantage to survive. There are many examples of decreases in bio-diversity in such areas. For example, Purple loosestrife (*Lythrum salicaria*) changed the ecology of wetlands by reducing the abundance of native plants and endangering several species of ducks and a species of turtle that depend on the native plants. Clearly, a primary threat to bio-diversity is the spread of human activity into once pristine areas.

Land clearing and human habitation put significant pressure on local species and disturbed habitat is often prone to invasions that can have adverse effects on local ecosystems, changing ecosystem functions. A species of wetland plant known as in Hawaii (the indigenous, *Bacopa monnieri*) is regarded as a pest species in artificially manipulated waterbird refuges because it quickly covers shallow mudflats established for endangered Hawaiian stilt (*Himantopus mexicanus knudseni*), making these undesirable feeding areas for the birds. Sometimes, multiple successive introductions of different non-native species can have interactive effects, where the introduction of a second non-native species can enable the first invasive species to flourish.

Examples of this are the introductions of the amethyst gem clam (*Gemma gemma*) and the European green crab (*Carcinus maenas*). The gem clam was introduced into California's Bodega Harbor from the East Coast of the United States a century ago. It had been found in small quantities in the harbor but had never displaced the native clam species (*Nutricola* spp.). In the mid 1990s, the introduction of the European green crab, found to prey preferentially on the native clams, resulted in a decline of the native clams and an increase of the introduced clam populations (Grosholz, 2005).

Invasive plants can arise from human clearing (such as slash-and-burn) or cattle grazing actions, such that the altered land is more hospitable to the invasive species than the original plant palette. The invasive plant can even be a pre-existing species that has attained dominance from the disturbance. For example, in the Waterberg region of South Africa, cattle grazing over the past six centuries has allowed invasive scrub and small trees to displace much of the original grassland, resulting in a massive reduction in forage for native bovids and other grazers. Since the 1970s large scale efforts have been underway to reduce invasives; partial success has led to re-establishment of many species that had dwindled or left the region. Examples of these species are giraffe, Blue Wildebeest, impala, kudu and White Rhino.

Invasive species can change the functions of ecosystems. For example invasive plants can alter the fire regime (cheatgrass, *Bromus tectorum*), nutrient cycling (smooth cordgrass *Spartina alterniflora*), and hydrology (*Tamarisk*) in native ecosystems (Mack *et al.* 2000). Invasive species that are closely related with rare native species have the potential to hybridize with native species. Harmful effects of hybridization have lead to a decline and even extinction of native species (Hawkes *et al.* 2005, Rhymer and Simberloff 1998). For example, hybridization with introduced cordgrass, *Spartina alterniflora*, threatens the existence of California cordgrass (*Spartina foliosa*) in San Francisco Bay (Ayres et al 2004).

Economic Impacts

Economic costs due to invasive species can be separated into direct costs due to production loss in agriculture and forestry, and management costs of invasive species. Estimated damage and control cost of invasive species in the U.S. alone amount to more than $138 billion annually (Pimentel *et al.* 2005). In addition to these costs, economic losses can occur due to loss from recreational and tourism revenues (Simberloff 2001).

Economic costs of invasions, when calculated as production loss and management costs, are low because they do not usually consider environmental damages. If monetary values could be assigned to the extinction of species, loss in bio-diversity, and loss of ecosystem services, costs from impacts of invasive species would drastically increase (Pimentel *et al.* 2005). The following examples from different sectors of the economy demonstrate the impact of biological invasions.

Agriculture

Agricultural weeds cause an overall reduction in yield. Most weed species are accidental introductions with crop seeds and imported plant material. Many introduced weeds in pastures compete with native forage plants, are toxic (*e.g.*, leafy spurge, *Euphorbia esula*) to cattle or non palatable due to thorns and spines (*e.g.*, yellow star thistle, *Centaurea solstitialis*). Forage loss due to invasive weeds on pastures amounts to nearly $1 billion in the U.S. alone (Pimentel *et al.* 2005). A decline in pollinator services and loss of fruit production has been observed due to the infection of honey bees (*Apis mellifera* another invasive species to the Americas) by the invasive varroa mite. Introduced rodents (rats, *Rattus rattus* and *R. norvegicus*) have become serious pests on farms destroying stored grains (Pimentel *et al.* 2005).

Forestry

The unintentional introduction of forest pest species and plant pathogens can change forest ecology and negatively impact timber industry. The Asian long-horned beetle (*Anoplophora glabripennis*) was first introduced into the U.S. in 1996 and is expected to infect and damage millions of acres of hardwood trees. Thirty million dollars have already been spent in attempts to eradicate this pest and protect millions of trees in the affected regions (Pimentel *et al.* 2005).

The woolly adelgid inflicts damage on old growth spruce fir forests and negatively impacts the Christmas tree industry (Forest Pests: Insects, Diseases & Other Damage Agents 2005). The chestnut blight fungus (*Cryphonectria parasitica*) and Dutch elm disease (*Ophiostoma novo-ulmi*) are two plant pathogens with serious impacts on forest health.

Tourism and Recreation

Invasive species can have impacts on recreational activities such as fishing, hunting, hiking, wildlife viewing, and water-based recreation. They negatively affect a wide array of environmental attributes that are important to support recreation, including but not limited to water quality and quantity, plant and animal diversity, and species abundance (Eiswerth 2005). Many invasive species have thorns and spikes than can prevent easy access to hiking trails. Aquatic invasive species, such as hydrilla (*Hydrilla verticillata*) and Eurasian watermilfoil (*Myriophyllum spicatum*), affect water-based recreation by impeding human access, interfering with the operation of boats and fishing lines, lowering water quality, and negatively altering aquatic ecosystems, including the abundance and diversity of fishes.

Health Impacts

An increasing threat of exotic diseases exists due to increased transportation and encroachment of humans into previously remote ecosystems that can lead to new associations between a disease and a human host (*e.g.*, AIDS virus in human host; Pimentel *et al.* 2005). Introduced birds (*e.g.* pigeons), rodents and insects (*e.g.* mosquitoes, fleas, lice and tsetse fly) can serve as vectors and reservoirs of human diseases. Throughout recorded history epidemics of human diseases such as malaria, yellow fever, typhus, and plague have been associated with these vectors (Elton 1958). A recent example of an introduced disease is the spread of the West Nile virus across North America resulting in human deaths and in the deaths of many birds, mammals, and reptiles (Lanciotti 1999). Waterborne disease agents, such as Cholera bacteria (*Vibrio cholerae*), and causative agents of harmful algal blooms are often transported via ballast water (Hallegraeff 1998). The full range of impacts of invasive species and their control goes beyond immediate effects and can have long term public health implications. For instance, pesticides applied to treat a particular pest species could pollute soil and surface water (Pimentel *et al.* 2005).

The Threat to Global Bio-diversity

The impact on global bio-diversity of human introduction of non native species that have subsequently become invasive is subjective. Climate change and the movement of the continents through the ages have created divisions and changes to species over the long history of this planet. Limited information on the circumstances and impact at the time makes it difficult to directly compare this to the advent of international travel for people and goods which has made the introduction of new species increasingly easy, and the direct ecological impacts are now far more evident and measurable.

Historically the deliberate introduction of non native species has been done with little or no consideration of the impact outside of having a favoured animal, fish, or plant available locally, or perhaps an ill-conceived attempt to control a native pest. In areas with highly endemic, specialised and isolated flora and fauna such as Australia, New Zealand, Madagascar, the Hawaiian Archipelago, and the Galapagos Islands, introduced species that successfully establish themselves in habitats utilized by natives compete for limited resources or prey on the native species, some of which are unable to adapt to the more competitive environment and gradually die out.

As more adaptable and generalized species are introduced to environments impacted adversely by human activities, native species are put at a disadvantage to survive in what previously was a unique, balanced ecosystem. There are many examples of decreases in bio-diversity in those areas. One of the primary threats to bio-diversity is the spread of humanity into what were once isolated areas with land clearing and habitation putting significant pressure on local species. Agriculture, livestock and fishing can also introduce changes to local populations of indigenous species which may result in a previously innocuous native species becoming a pest due to a reduction of natural predators.

Control of Invasive Species

The control of invasive species can involve their eradication or their containment within a specified area. In both cases, the goal is to prevent further spread to un-invaded systems (National Biological Information Infrastructure 2004). This type of management can be implemented at several scales, from a homeowner working in his or her own backyard to large government agencies taking a national approach. The decision to eradicate a species versus contain it can depend on several factors, including, but not limited to, the type of habitat, characteristics of the organism, the spatial dimensions of the spread, time available to dedicate to control, and cost. These factors also play a role in determining which specific control technique(s) to utilize.

Mechanical Control

Mechanical control involves the removal of invasive species by hand or with machines. Often, these methods are effective in controlling small populations and can be target specific, minimizing harm to non-invasive plants and animals (The Nature Conservancy 2005). Mechanical control, however, is extremely labour intensive and requires a large time investment, as treatments must often be applied several times to ensure success (The Nature Conservancy 2005). Commonly implemented control methods for plants include hand pulling, mowing, girdling, and burning (The Nature Conservancy 2005). For invasive animal control, techniques such as hunting, trapping, and the construction of physical barriers like fences or nets, are used (Hengeveld 1989).

Chemical Control

Chemical compounds can be used to prevent the spread of invasive species. This method of control can be very effective in both large and small areas, but is often criticized due to the possible contamination of land and water resources and a lack of target specificity that can result in the killing of desirable plant and animal species (National Park Service 2004). Also, the target species may develop a resistance to the chemicals over time, rendering this method ineffective. Herbicides are chemicals used to control invasive plants and, depending on the target species, can be applied directly to a plant, in the soil at a plant's base, or even to the soil before seeds develop (The Nature Conservancy 2005). For animals, other pesticides are used to restrict growth and reproduction or to kill invasive insect pests (National Biological Information Infrastructure 2004). Another form of chemical control is the use of attractant pheromones to lure mate-seeking insects into traps (Myers *et al.* 2000).

Biological Control

Biological control involves the release of a specific species to restrict the spread of the invasive species. With the proper research, this method of control can be both environmentally safe and successful. However, it can be ineffective if the released species do not survive or if their impact on the invasive species is not as great as predicted (The Nature Conservancy 2005). Also, the species chosen for release is not always a native organism, increasing the possibility of even more invasive species.

Predatory insects, called weed feeders, can be released to control invasive plants. Similarly, plants can be infected by disease causing organisms, such as fungi, bacteria, and viruses, killing them or reducing their reproductive output (Cornell University 2005). Invasive animals can be controlled with the release of predatory or parasitic organisms (especially in the case of invasive insects) or with the transmission of diseases in a similar manner as with plants (Cornell University 2005). Additionally, sterile insect or fish males of the invasive species can be released so that after mating, a female will lay "dead" eggs or eggs that will develop into sterile adults (Myers *et al.* 2000).

Prevention

Preventing the establishment of invasive species is always the best method of control (The Nature Conservancy 2005). Stopping harmful species at this stage can be difficult. Many governments try to limit the entry of invasive species into their lands with thorough inspections of international shipments, customs checks, and proper quarantine regulations. The creation of list of safe and potentially harmful species can be helpful in regulation. The general public can also participate in invasive species prevention by educating themselves about invasive species and by making informed decisions.

Seed

A seed is a small embryonic plant enclosed in a covering called the seed coat, usually with some stored food. It is the product of the ripened ovule of gymnosperm and angiosperm plants which occurs after fertilization and some growth within the motherplant. The formation of the seed completes the process of reproduction in seed plants (started with the development of flowers and pollination), with the embryo developed from the zygote and the seed coat from the integuments of the ovule.

This process starts with double fertilization in angiosperms and it involves the fusion of the egg and sperm nuclei into a zygote. The second part of this process is the fusion of the polar nuclei with a second sperm cell nucleus, thus forming a primary endosperm. Right after fertilization the zygote is mostly inactive but the primary endosperm divides rapidly to form the endosperm tissue. This tissue becomes the food that the young plant will consume until the roots have developed after germination or it develops into a hard seed coat. The seed, which is an embryo with two points of growth (one of which forms the stems the other the roots) is enclosed in a seed coat with some food reserves. In gymnosperms the two sperm cells transferred from the pollen do not develop seed by double fertilization but instead only one sperm fertilizes the egg while the other is not used. The seed is composed of the embryo (the result of fertilization) and tissue from the mother plant, which also form a cone around the seed in coniferous plants like Pine and Spruce.

The new seed is formed in plant structures called fruits in angiosperms.

Seed also has a general meaning that predates the above-anything that can be sown *i.e.* "seed" potatoes, "seeds" of corn or sunflower "seeds". In the case of sunflower and corn "seeds", what is sown is the seed enclosed in a shell or hull, and the potato is a root or stem depending on what is being sowed.

Seeds have been an important development in the reproduction and spread of flowering plants, relative to more primitive plants like mosses, ferns and liverworts, which do not have seeds and use other means to propagate themselves. This can be seen by the success of seed plants (both gymnosperms and angiosperms) in dominating biological niches on land, from forests to grasslands both in hot and cold climates.

Seed Structure

A seed contains the embryo from which a new plant will grow under proper conditions. Seeds also usually contain a supply of stored energy and is wrapped in the seed coat or testa. Seeds are very diverse in size. The dust-like orchid seeds are the smallest with about one million seeds per gram. Embryotic seeds have immature embryos and no significant energy reserves. They are myco-heterotrophs, depending on mycorrhizal fungi for nutrition during germination. At over 20 kg, the largest seed is the coco de mer.

The embryo has one cotyledon or seed leaf in monocotyledons, two cotyledons in almost all dicotyledons and two or more in gymnosperms. The radicle is the embryonic root. The plumule is the embryonic shoot. The embryonic stem above the point of attachment of the cotyledon(s) is the epicotyl. The embryonic stem below the point of attachment is the hypocotyl.

In angiosperms, the stored food begins as a tissue called the endosperm, which is derived from the parent plant via double fertilization. The usually triploid endosperm is rich in oil or starch and protein. In gymnosperms, such as conifers, the food storage tissue is part of the female gametophyte, a haploid tissue.

In some species, the embryo is embedded in the endosperm or female gametophyte, which the seedling will use upon germination. In others, the endosperm is absorbed by the embryo as the latter grows within the developing seed, and the cotyledons of the embryo become filled with this stored food. At maturity, seeds of these species have no endosperm and are termed exalbuminous seeds. Some exalbuminous seeds are bean, pea, oak, walnut, squash, sunflower, and radish. Seeds with an endosperm at maturity are termed albuminous seeds. Most monocots (*e.g.* grasses and palms) and many dicots (*e.g.* brazil nut and castor bean) have albuminous seeds. All gymnosperm seeds are albuminous.

The seed coat develops from the tissue, the integument, originally surrounding the ovule. The seed coat in the mature seed can be a paper-thin layer (*e.g.* peanut) or something more substantial (*e.g.* thick and hard in honey locust and coconut).

The seed coat helps protect the embryo from mechanical injury and from drying out.

The seeds of angiosperms are contained in a hard or fleshy (or with layers of both) structure called a fruit. Gymnosperm seeds begin their development "naked" on the bracts of cones, although the seeds do become covered by the cone scales as they develop. An example of a hard fruit layer surrounding the actual seed is that of the so-called *stone* fruits (such as the peach).

Some seeds have an appendage on the seed coat such an aril (as in yew and nutmeg) or an elaiosome (as in Corydalis) or hairs (as in cotton). The hilum is the scar on the seed coat where the seed was attached to the ovary wall by the funiculus.

In order for the seed coat to split, the embryo must imbibe (soak up water), which causes it to swell, splitting the seed coat. However, the nature of the seed coat determines how rapidly water can penetrate and subsequently initiate germination. For seeds with a very thick coat, scarification of the seed coat may be necessary before water can reach the embryo. Examples of scarification include: gnawing by animals, freezing and thawing, battering on rocks in a stream bed, or passing through an animal's digestive tract. In the latter case, the seed coat protects the seed from digestion, while perhaps weakening the seed coat such that the embryo is ready to sprout when it gets deposited (along with a bit of fertilizer) far from the parent plant. In species with thin seed coats, light may be able to penetrate into the dormant embryo. The presence of light or the absence of light may trigger the germination process, inhibiting germination in some seeds buried too deeply or in others not buried in the soil. Abscisic acid is usually the growth inhibitor in seeds.

Seed Functions

Seeds protect and nourish the embryo or baby plant. Seeds usually give a seedling a faster start than a sporling from a spore gets because of the larger food reserves in the seed.

Unlike animals, plants are limited in their ability to seek out favorable conditions for life and growth. As a consequence, plants have evolved many ways to disperse their population through their seeds. A seed must somehow "arrive" at a location and be there at a time favorable for germination and growth. Seed dispersal is often attributed mainly to fruits, however many seeds aid in their own dispersal, for example:

- Many seeds (*e.g.* maple, pine) have a wing that aids in wind dispersal.
- The dustlike seeds of orchids are carried efficiently by the wind.
- Some seeds, (*e.g.* dandelion, milkweed, poplar) have hairs that aid in wind dispersal.
- Seeds with a fleshy covering (*e.g.* apple, cherry, juniper) are eaten by animals (birds, mammals) which then disperse these seeds in their droppings.
- Seeds (nuts) which are an attractive long-term storable food resource for animals (*e.g.* acorns, hazelnut, walnut); the seeds are stored some distance from the parent plant, and some escape being eaten if the animal stores more than it needs, forgets where it has stored the seed, or dies.
- Seeds (burrs) with barbs or hooks (*e.g.* acaena, burdock, dock) which attach to animal fur or feathers, and then drop off later.
- Seeds of some mangroves are viviparous, they begin to germinate while still attached to the parent. The large, heavy root allows the seed to penetrate into the ground when it falls.
- Some seeds have appendages called elaiosomes, *e.g.* bloodroot, trilliums and Acacias. Elaiosomes provide food for ants, which usually disperse such seeds.
- Some plants, such as *Mucuna* and *Dioclea*, produce buoyant seeds termed sea-beans or drift seeds because they float in rivers to the oceans and wash up on beaches.

For annuals, seeds are a way for the species to survive dry or cold seasons. Ephemeral plants are usually annuals that can go from seed to seed in as few as six weeks.

One important function of most seeds is delaying germination to allow time for dispersal and to prevent all seeds from germinating at once when conditions are favorable. Staggering germination prevents all seeds from germinating at once and being wiped out by bad weather or herbivores. Seed dormancy is defined as a seed failing to germinate under environmental conditions optimal for germination. It is often confused with seed quiescence, which is a seed failing to germinate because environmental conditions are inappropriate for germination. Many cultivated seeds lack dormancy but do not germinate in seed packets simply because there is insufficient moisture.

Origin and Evolution

The origin of seed plants is a problem that still remains unsolved. However, more and more data tends to place this origin in the middle Devonian. The description in 2004 of the proto-seed *Runcaria heinzelinii* in the Givetian of Belgium is an indication of that ancient origin of seed-plants.

The first "true" seeds are described from the upper Devonian, which is probably the theatre of their true first evolutionary radiation. The seed plants progressively became one of the major elements of nearly all ecosystems.

Economic Importance

Many seeds are edible. The majority of human calories comes from seeds, especially from cereals, legumes and nuts. Seeds also provide most cooking oils, many beverages and spices and some important food additives.

Seeds are used to propagate many crops such as cereals, legumes, forest trees, turfgrasses and pasture grasses.

Some seeds are also poisonous. One of the deadliest poisons, ricin, comes from seeds of the castor bean. Another seed poison is strychnine. Other poisonous seeds are those of yew, wisteria, apple, horse-chestnut and peach.

The world's most important clothing fibre grows attached to cotton seed. Other seed fibres are from kapok and milkweed.

Many important non-food oils are extracted from seeds. Linseed oil is used in paints. Oil from jojoba and crambe are similar to whale oil.

Seeds are the source of some medicines including castor oil, tea tree oil and the discredited cancer drug, Laetrile.

Many seeds have been used as beads in necklaces and rosaries including Job's tears, Chinaberry and rosary pea. However, the latter two are also poisonous.

Other seed uses include:

- Seeds once used as weights for balances.
- Seeds used as toys by children, such as for the game conker.
- Resin from *Clusia rosea* seeds used to caulk boats.
- Nematicide from milkweed seeds.
- Cottonseed meal used as animal feed and fertilizer.
- Many seeds used as birdseed.

Oldest Viable Seeds

- The oldest carbon-14-dated seed that has grown into a viable plant was a Judean date palm seed about 2,000 years old, recovered from excavations at Herod the Great's palace on Masada in Israel. It was germinated in 2005.

Seed Packets and Seed Information

Generally, seed packets labels includes:

- Common plant name and the botanical name (in parentheses).
- Space and deep: how deep to place the seeds in the soil, space between plants (from one row to the other one and from one plant to the other one in the same row).
- Height: approximate height the plant will reach when mature.
- Soil: type of soil the plant prefers.
- Water: It can indicate "keep the soil lightly damp", "bottom water the plant", "drench the soil with water", "daily misting of water" and "almost dry out before re-watering".

- Sun: full direct sunlight, partial sun, diffused sunlight, or grows well in the shade.
- Door: if the plant is best suited for growing Indoor, Outdoor or Both.
- Live: Perennial or annual.
- Planting, germination and harvest period: This information can be indicated by months or quarters of the year.
- Special requirements, if necessary.

This information can be represented graphically.

Experiments with Hybrids of Other Species of Plants

It must be the object of further experiments to ascertain whether the law of development discovered for Pisum applies also to the hybrids of other plants. To this end several experiments were recently commenced. Two minor experiments with species of Phaseolus have been completed, and may be here mentioned.

An experiment with Phaseolus vulgaris and Phaseolus nanus gave results in perfect agreement. Ph. nanus had together with the dwarf axis, simply inflated, green pods. Ph. vulgaris had, on the other hand, an axis 10 ft. to 12 ft. high, and yellow coloured pods, constricted when ripe. The ratios of the numbers in which the different forms appeared in the separate generations were the same as with Pisum. Also the development of the constant combinations resulted according to the law of simple combination of characters, exactly as in the case of Pisum. There were obtained:

Constant combinations:	Axis:	Color of the unripe pods:	Form of the ripe pods:
1	long	green	inflated
2	long	green	constricted
3	long	yellow	inflated
4	long	yellow	constricted
5	short	green	inflated
6	short	green	constricted
7	short	yellow	inflated
8	short	yellow	constricted

The green colour of the pod, the inflated forms, and the long axis were, as in Pisum, dominant characters.

Another experiment with two very different species of Phaseolus had only a partial result. Phaseolus nanus L, served as *seed parent*, a perfectly constant species, with white flowers in short recemes and small white seeds in straight, inflated, smooth pods; as *pollen parent* was used Ph. multiflorus, W, with tall winding stem, purple-red flowers in very long recemes, rough, sickle-shaped crooked pods, and large seeds which bore black flecks and splashes on a peach-blood-red ground.

The hybrids had the greatest similarity to the pollen parent, but the flowers appeared less intensely coloured. Their fertility was very limited; from 17 plants, which together developed many hundreds of flowers, only 49 seeds in all were obtained. These were of medium size, and were flecked and splashed similarly to those of Ph. multiflorus, while the ground colour was not materially different.

The next year 44 plants were raised from these seeds, of which only 31 reached the flowering stage. The characters of Ph. nanus, which had been altogether latent in the hybrids, reappeared in various combinations; their ratio, however, with relation to the dominant plants was necessarily very fluctuating owing to the small number of trial plants. With certain characters, as in those of the axis and the form of pod, it was, however, as in the case of Pisum, almost exactly 1:3.

Insignificant as the results of this experiment may be as regards the determination of the relative numbers in which the various forms appeared, it presents, on the other hand, the phenomenon of a *remarkable change of colour* in the flowers and seed of the hybrids.

In Pisum it is known that the characters of the flower- and seed-colour present themselves unchanged in the first and further generations, and that the offspring of the hybrids display exclusively the one or the other of the characters of the original stocks. It is otherwise in the experiment we are considering.

The white flowers and the seed-colour of Ph. nanus appeared, it is true, at once in the first generation in one fairly fertile example, but the remaining 30 plants developed flower-colours which were of various grades of purple-red to pale

appeared once among 31 plants of the first generation. This colouring appears only once in the series, and could therefore also only be developed once in the average in each 16, and with three colour characters only once even in 64 plants.

It must, nevertheless, not be forgotten that the explanation here attempted is based on a mere hypothesis, only supported by the very imperfect result of the experiment just described. It would, however, be well worth while to follow up the development of colour in hybrids by similar experiments, since it is probable that in this way we might learn the significance of the extraordinary variety in the *colouring of our ornamental flowers*.

So far, little at present is known with certainty beyond the fact that the colour of the flowers in most ornamental plants is an extremely variable character. The opinion has often been expressed that the stability of the species is greatly disturbed or entirely upset by cultivation, and consequently there is an inclination to regard the development of cultivated forms as a matter of chance devoid of rules; the colouring of ornamental plants is indeed usually cited as an example of great instability. It is, however, not clear why the simple transference into garden soil should result in such a thorough and persistent revolution in the plant organism.

No one will seriously maintain that in the open country the development of plants is ruled by other laws than in the garden bed. Here, as there, changes of type must take place if the conditions of life be altered, and the species possesses the capacity of fitting itself to its new environment. It is willingly granted that by cultivation the origination of new varieties is favoured, and that by man's labour many varieties are acquired which, under natural conditions, would be lost; but nothing justifies the assumption that the tendency to formation of varieties is so extraordinarily increased that the species speedily lose all stability, and their offspring diverge into an endless series of extremely variable forms. Were the change in the conditions the sole cause of variability we might expect that those cultivated plants which are grown for centuries under almost identical conditions would again attain constancy.

This, as is well known, is not the case since it is precisely under such circumstances that not only the most varied but also the most variable forms are found. It is only the Leguminosae, like Pisum, Phaseolus, Lens, whose organs of fertilization are protected by the keel, which constitute a noteworthy exception. Even here there have arisen numerous varieties during a cultural period of more than 1000 years under most various conditions; these maintain, however, under unchanging environments a stability as great as that of species growing wild.

It is more than probable that as regards the variability of cultivated plants there exists a factor which so far has received little attention. Various experiments force us to the conclusion that our cultivated plants, with few exceptions, are *members of various hybrid series*, whose further development in conformity with law is varied and interrupted by frequent crossings *inter se*.

The circumstance must not be overlooked that cultivated plants are mostly grown in great numbers and close together, affording the most favorable conditions for reciprocal fertilization between the varieties present and species itself. The probability of this is supported by the fact that among the great array of variable forms solitary examples are always found, which in one character or another remain constant, if only foreign influence be carefully excluded.

These forms behave precisely as do those which are known to be members of the compound hybrid series. Also with the most susceptible of all characters, that of colour, it cannot escape the careful observer that in the separate forms the inclination to vary is displayed in very different degrees. Among plants which arise from *one* spontaneous fertilization there are often some who offspring vary widely in the constitution and arrangement of the colours, while that of others shows little deviation, and among a greater number solitary examples occur which transmit the colour of the flowers unchanged to their offspring.

The cultivated species of Dianthus afford an instructive example of this. A white-flowered example of Dianthus caryophyllus, which itself was derived from a white-flowered variety, was shut up during its blooming period in a greenhouse; the numerous seeds obtained therefrom yielded plants entirely white-flowered like itself. A similar result was obtained from a sub-species, with red flowers somewhat flushed with violet, and one with flowers white, striped with red. Many others, on the other hand, which were similarly protected, yielded progeny which were more or less variously coloured and marked.

Whoever studies the coloration which results in ornamental plants from similar fertilization can hardly escape the conviction that here also the development follows a definite law which possibly finds its expression *in the combination of several independent colour characters*.

Bibliography

Adams H.: *Hot Pepper Improvement*, St. Michael, CARDI, 1997.

Agarwal P. K.: *Improvement of Citrus*, New Delhi, Malhotra Publishing House, 1993.

Alexander M. P. and Ganeshan S.: *Pollen Storage*, New Delhi, Malhotra Publishing House, 1993.

Andrews J.: *The Domesticated Capsicums*, University of Texas Press. 1995.

Andrews L.: *Citrus Production - Orange*, St. Augustine, Trinidad and Tobago, 1990

————: *Production of Minor Crops Sapodilla, Tamarind, Guava and Genip*, Port of Spain, Trinidad and Tobago, 1996.

Arguello D. S.: *Production, Post-harvest Management and Exportation of Tropical Fruits in Costa Rica*, Wageningen, Technical Centre for Agricultural and Rural Cooperation, 1992.

Ashworth S.: *Seed to Seed*, Decorah, Seed Savers Publications, 1991.

Barbeau G.: *Tropical Fruits in Nicaragua*, Managua, Nicaragua Ministerio de Desarrollo Agropecuario, Agraria, 1990.

Batlle I. and Tous J.: *Carob Tree (Ceratonia siliqua L.)*. Rome, International Plant Genetic Resources Institute, 1997.

Berger, P.L.: *Pyramids of Sacrifice: Political Ethics and Social Change*, New York, Basic Books, 1974.

————: *Pyramids of Sacrifice: Political Ethics and Social Change*, New York, Basic Books, 1974.

Brooks, D.: *Water: Local-Level Management*, Ottawa, International Development Research Centre, 2002.

Bunnik J. S. C.: *Fresh Fruits and Vegetable: a Survey on the Netherlands and other Major Markets in the European*

Community, Netherlands, Centre for the Promotion of Imports from Developing Countries, 1990.

Burton W.G.: *The Potato*, Holland, H. Veenman & Zonen N.V., 1966.

Cao Van P.: *An Integrated Approach for the Production and Processing of Minor Fruits in Martinique (FWI)*, Port of Spain, Trinidad and Tobago, 1996.

Cardona M. J. G., Carvajal S. L. B., Salinas D. G. C. and Isaza R. G. B. : *Proceedings of the International Seminar on Plantain Production, Quindio, Colombia,* Quindio, Corporation Colombiana de Investigation Agropecuaria, 1998.

Chadha K. L. and Pareek O. P.: *Advances in Horticulture: Fruit Crops,* New Delhi, Malhotra Publishing House, 1993.

Chambers, R.: *Rural Development: Putting the Last First*, London, Longman, 1983.

Collymore L.: *Fruit Production in Barbados*, Port of Spain, Trinidad and Tobago, 1996.

Coste R.: *Coffee: the Plant and the Product*, London, MacMillan, 1992.

Crucefix D.: *Avocado Variety Selection for Export Development,* Roseau, CARDI, 1996.

Cull Brian & Lindsay Pax: *Fruit Growing in Warm Climates for Commercial Growers & Home Gardeners*, Australia, Reed Books, 1995.

Currah L. and Proctor F. J.: *Onions in Tropical Regions*, Kent, Natural Resources Institute, 1990.

Daniells J.: *Illustrated Guide to the Identification of Banana varieties in the South Pacific*, Canberra, ACIAR, 1995.

Degras L.: *Yam: a Tropical Root Crop*, Wageningen, CTA/ MacMillan, 1993.

Dhatt A. S. and Singh Z.: *Propagation and Rootstocks of Citrus*, New Delhi, Malhotra Publishers, 1993.

Diederichsen A.: *Coriander (Coriandrum sativum L.)*. Rome, International Plant Genetic Resources Institute, 1996.

Doijode S. D.: *Seed Germination in Fruits*, New Delhi, Malhotra Publishers, 1993.

Dudley, E.: *The Critical Villager: Beyond Community Participation*, London, Routledge, 1993.

Featherly H. I.: *Taxonomic Terminology of the Higher Plants*, USA, Iowa State College Press, 1954.

Ferentinos L.: *Proceeding of the Sustainable Taro Culture for the Pacific Conference*, Honolulu, HITAHR, 1993.

Forde S: *Proceedings of CARDI / CTA Workshop on Marketability of Caribbean Minor Fruits*, Port of Spain, Trinidad and Tobago, 1996.

Georges, S.: *The Debt Boomerang: How Third World Debt Harms Us All*, Boulder, Westview Press, 1992.

Glowinski Louis: *The Complete Book of Fruit Growing in Australia*, Australia, Lothian Publishing Company Pty. Ltd., 1991.

Godden G.: *Growing Citrus Trees*, Australia, Lothian Publishing Company Pvt. Ltd., 1988.

Gowen S.: *Banana and Plantains*, London, Chapman Hall, 1996.

Green S.K. and Kim J. S.: *Sources of Resistance to Viruses of Pepper (Capsicum spp.): a Catalogue*, Taipei, AVRDC, 1994.

Gunjate R. T. and Tawde A. B.: *Propagation and Rootstocks of Mango*, New Delhi, Malhotra, 1993.

Hartley W.: *A Checklist of Economic Plants in Australia*, Melbourne, C.S.I.R.O., 1979.

Harwood, R. R.: *Marshalling Technology for Development: Proceedings of a Symposium*, Washington, DC, National Academy Press, 1995.

Herklots G. A. C.: *Vegetables in South East Asia*, London, George Allen & Unwin Ltd., 1972.

Hessayon D. G. Dr.: *The Vegetable Expert*, England, PBI, Publications, 1985.

Huaman Z.: *Descriptors for Sweet Potato*, Rome, International Board for Plant Genetic Resources, 1991.

Hurst Jacqui & Rutherford Lyn.: *A Gourmet's Book of Mushrooms & Truffles*, Sydney, Golden Press Pvt. Ltd., 1991.

Jacquat Christiane: *Plants from the Markets of Thailand*, Bangkok, Duang Kamol, 1990.

Jeffers P.: *Evaluation of Four Onion Varieties in Montserrat*, Plymouth, CARDI, 1992.

Kenridge K. C. and Hardy B.: *Biology and Agronomy of Forage Arachis*, Cali, International Centre for Tropical Agriculture, 1994.

Kroll R.: *Cut Flowers*, Wageningen, CTA, 1995.

Kunelius T.: *Annual Ryegrasses in Atlantic Canada*, Ottawa, Agriculture Canada, 1991.

Lofgren, H., Richards, A.: *Food Security, Poverty, and Economic Policy in the Middle East and North Africa*, Washington, DC, International Food Policy Research Institute, 2003.

Mabberley D. J.: *The Plant-Book : a Portable Dictionary of the Vascular Plants*, Cambridge, Cambridge University Press, 1997.

Madulid Domingo A.: *A Pictorial Cyclopedia of Philippine Ornamental Plants*, Philippines, Makati Metro Manila, 1995.

Malins A.: *Postharvest Handling of Pineapple and Mango*, Port of Spain, Trinidad and Tobago, 1992.

Mannetje, L. T. & Jones, R. M.: *Plant Resources of South-East Asia,* Wageningen, Pudoc Scientific Publishers, 1992.

Matsuoka H.: *Cultivation of Panicum Genetic Resources for Evaluation of Characteristics*, Tokyo, JICA, 1997.

Mc Donald F.: *Plant Tissue Culture Manual for Yam, Cassava, Sweet Potato, Dasheen (taro) and Tannia (cocoyam)*, Roseau, Dominica and Cave Hill, 1993.

Mellor, J. W.: *The New Economics of Growth*, Ithaca, Cornell University Press, 1976.

Miller William: *Dictionary of English Plant Names*, London, John Murray, 1884.

Mitra S.: *Postharvest Physiology and Storage of Tropical and Subtropical Fruits*, Oxon, CABI, 1997.

Morris M. L. : *Maize Seed Industries in Developing Countries*, Colorado, Lynne Rienner, 1998.

Morton Julia F.: *Fruits of Warm Climates*, Miami, Julia F. Morton Publisher, 1987.

Murali T. P. and Duncan E. J.: *In Vitro Propagation of Banana through Aseptic Manipulation of Male Inflorescences*, Port of Spain, NIHERST, 1990.

Nabhan G. P.: *Wild Phaseolus Ecogeography in the Sierra Madre Occidental, Mexico: Areographic Techniques for Targeting and Conserving Species Diversity*, Rome, International Board for Plant Genetic Resources, 1990.

Nijdam J. & De Jong A.: *Elsevier's Dictionary of Horticulture in Nine Languages*, Elsevier Scientific Publishing Co. Amsterdam; New York.

Oka H. I.: *Origin of Cultivated Rice*, Elsevier, Japan Scientific Societies Press, 1988.

Oldham P.: *Cost of Production of Major Tree Crops in Dominica*, Roseau, Ministry of Agriculture, 1991.

Ou Yang Jue Ya: *Comparative Study of Mandarin and Cantonese*, Beijing, China Social Sciences Publishing House, 1993.

Percival John: *The Wheat Plant - A Monograph*, London, Duckworth & Co., 1921.

Pilgrim R.: *Post Harvest Handling of Minor Exotics*, St. George's, Grenada, 1996.

Ragone D.: *Breadfruit: Artocarpus Altilis (Parkinson) Fosberg*, Rome, International Plant Genetic Resources Institute, 1997.

Rehm Sigmund: *Multilingual Dictionary of Agronomic Plants*, Boston, Kluwer Academic Publishers, 1994.

Roy S. K.: *Research on Multipurpose Tree Species in Asia: In Vitro Clonal Propagation of Artocarpus Heterophyllus*, Bangkok, Winrock International Institute for Agricultural Development, 1991.

Singh H. P. and Chadha K. L.: *Genetic Resources of Citrus*, New Delhi, Malhotra Publishing House, 1993.

Sinha G. C., Reddy Y. T. N. and Singh G.: *Propagation and Rootstocks in Guava*, New Delhi, Malhotra Publishers, 1993.

Sperling L. and Berkowitz P. *Partners in selection: bean breeders and women bean experts in Rwanda*. Washington, USA: CGIAR, 1994.

Sreekumar V., Indrasenan G and Mammen G.: *Studies of the Quantitative and Qualitative Attributes of Ginger Cultivars*, Calicut, Kasaragod, 1990.

Stover R. H. and Simmonds N. W.: *Bananas*, United Kingdom: Longman Scientific and Technical, 1991.

Thavarasook C.: *Proceedings of the Mungbean Meeting 90, Chiang Mai, Thailand,* Bangkok, Tropical Agricultural Research Centre, 1991.

Thomas E.: *Fruit Production in St. Kitts and Nevis*, Port of Spain, IICA, 1996.

Vargas E. M., Macaya G., Baudoin J. P. and Rocha O. J.: *Variation in the Content of Phaseolin in Wild Populations of Lima Beans (Phaseolus lunatus L.) in the Central Valley of Costa Rica*. Plant Genetic Resources Newsletter, 2000.

Webb M.: *Weed Problems and their Control in Rice, Papaya, Citrus and Sugar cane: a Report to the Belize Plant Protection Service*, United Kingdom, Natural Resources Institute, 1993.

Whealy K.: *The Garden Seed Inventory*, Decorah, Seed Saver Publications, 1988.

Whitwell A.: *Dominica Orchard Crop Management and Research Project: Pest and Disease Management,* United Kingdom, Natural Resources Institute, 1991.

Woolfe Jennifer A.: *The Potato in the Human Diet*, Cambridge, Cambridge University Press, 1989.

Yoshida T.: *Cultivation of Citrus Genetic Resources for Evaluation of Characteristics*, Tokyo, JICA, 1996.

Index

A

Acer Nigrum, 27.
Agricultural Sciences, 141.
Allspice, 171, 172.
Aphids, 155, 156, 166, 170.
Arecanut, 161, 162, 175.
Ayurvedic, 136.

B

Beans, 139, 177, 178, 179, 180, 220.
Biodynamic Agriculture, 139, 142, 148.
Biodynamic Cultivation, 139, 148, 156.
Biodynamic Preparations, 143, 144, 154, 155.
Biological Diversity, 107, 108, 115, 116, 117, 118, 121, 123, 128, 130.
Butternut, 28, 30, 32.

C

Cardamom, 80, 186, 187, 189, 190, 191, 192.
Carrots, 139, 140, 141.
Cherry, 24, 25, 42, 55, 56, 220.
Climate, 58, 65, 75, 93, 100, 101, 102, 104, 141, 143, 144, 160, 164, 167, 169, 171, 173, 181, 183, 186, 200, 213.
Clove, 80, 171, 172, 182, 183, 184, 185.
Coconut, 161, 175, 188, 218.
Community, 33, 75, 85, 86, 87, 93, 114, 116, 125, 130, 208.
Conditions, 6, 7, 19, 30, 40, 42, 47, 57, 59, 65, 68, 69, 70, 83, 91, 116, 117, 134, 143, 153, 155, 160, 163, 164, 165, 167, 178, 183, 185, 190, 201, 202, 207, 218, 219, 221, 227, 228.
Conservation, 8, 10, 40, 41, 83, 90, 91, 92, 93, 97, 107, 108, 116, 197, 199.

D

Destruction, 60, 90, 91, 97, 149.
Development, 15, 76, 77, 92, 98, 111, 113, 114, 115, 116, 127, 128, 131, 133, 148, 150, 153, 154, 155, 168, 176, 177, 181, 191, 217, 219, 223, 225, 226, 227, 228, 229.
Distribution, 13, 22, 30, 71, 74, 84, 107, 118, 120, 126, 186.
Diversity, 1, 78, 83, 84, 85, 86, 87, 88, 90, 91, 92, 93, 95, 97, 101, 104, 105, 106, 107, 108, 111, 112, 114, 115, 116, 117, 118, 121, 123, 128, 130, 137, 198, 205, 208, 210, 211, 213, 214.

E

Ecosystem, 75, 82, 83, 84, 86, 87, 90, 93, 109, 114, 124, 125, 156, 193, 201, 203, 205, 206, 208, 210, 211, 214.
Environmental Protection, 115, 193.
Environments, 17, 55, 60, 62, 67, 91, 114, 148, 210, 214, 228.
Export, 127, 132, 179, 180.
Extraction, 178, 179, 188.

F

Fertilizers, 162, 165, 181, 184.
Flowers, 15, 21, 31, 33, 35, 39, 42, 46, 55, 56, 57, 59, 62, 63, 64, 67, 69, 71, 140, 141, 142, 144, 164, 170, 172, 176, 180, 186, 217, 224, 225, 226, 227, 228, 229.
Food, 17, 33, 35, 36, 37, 40, 48, 53, 57, 61, 64, 71, 80, 81, 92, 98, 100, 101, 111, 113, 114, 117, 119, 138, 139, 140, 141, 147, 149, 150, 151, 152, 160, 182, 186, 195, 200, 204, 207, 216, 217, 218, 219, 220, 221, 222.
Forestry, 108, 111, 114, 121, 129, 197, 211, 212.
Fruit, 15, 21, 22, 23, 24, 25, 26, 27, 28, 30, 31, 32, 33, 34, 35, 36, 37, 38, 39, 42, 43, 44, 47, 52, 56, 67, 70, 80, 145, 146, 160, 161, 163, 164, 170, 171, 172, 176, 183, 186, 201, 212, 219, 225.
Fungi, 5, 54, 92, 150, 151, 152, 216, 218.

G

Gardens, 52, 54, 55, 56, 59, 61, 62, 63, 64, 65, 66, 67, 68, 70, 139, 191, 193.
Garlic, 80, 166, 167, 168, 196.
Guidelines, 77, 108, 110, 119, 121, 125, 126, 127, 130, 131, 137.

H

Harvest, 60, 88, 90, 92, 154, 162, 166, 168, 169, 172, 183, 185, 191, 192, 223.
Harvesting, 135, 162, 166, 167, 168, 169, 171, 172, 177, 182, 184, 185, 188, 191.
Hawthorn, 36, 51, 59, 60.
Herbs, 135, 136, 137, 153, 154, 156.

I

Industry, 79, 115, 137, 172, 182, 183, 195, 212.
Infrastructure, 112, 132, 214, 215.
Insects, 5, 33, 59, 96, 156, 178, 200, 212, 213, 215, 216.
International Trade, 108, 109, 111, 127, 129.
Invasive Alien Species, 107, 108, 109, 110, 111, 112, 113, 114, 115, 116, 117, 118, 119, 120, 121, 122, 123, 124, 125, 126, 128, 129, 130, 131, 132, 133, 134.
Invasive Species, 114, 131, 193, 194, 197, 201, 202, 203, 206, 207, 208, 209, 210, 211, 212, 213, 214, 215, 216.
Irrigation, 161, 165, 167, 168, 169, 181, 184, 188, 190, 191.

J

Jams, 57, 160, 162.
Juglans Cinerea, 32.
Juglans Nigra, 28.

L

Leaf, 22, 28, 29, 30, 31, 32, 35, 38, 39, 43, 44, 47, 52, 55, 60, 67, 68, 70, 80, 141, 142, 143, 144, 145, 146, 155, 164, 166, 168, 169, 170, 172, 176, 186, 189, 218.

M

Mass Extinction, 94, 97, 99, 100, 101, 104, 106.
Medicinal Plants, 137, 138, 139, 140, 141, 142, 143, 144, 146, 147, 148, 149, 150, 153, 155, 156.
Medicine, 135, 136, 137.
Monitoring, 76, 77, 80, 109, 126.
Morphological Species, 6.

N

Native Plants, 197, 200, 203, 204, 210.
Natural Habitats, 91, 108, 112, 149, 150.
Nutmeg, 80, 160, 161, 162, 163, 171, 219.
Nutrients, 98, 150, 151, 152, 153, 156, 203.

O

Oils, 80, 81, 140, 141, 142, 155, 163, 221, 222.
Orange, 31, 32, 36, 37, 38, 40, 42, 45, 46, 48, 61, 63, 66.
Ornamental Plants, 196, 197, 227, 229.

P

Parameters, 85, 88, 89, 90, 92, 118.
Parks, 55, 56, 59, 61, 63, 65, 66, 68, 69, 170, 196.
Pesticides, 137, 138, 139, 150, 155, 213, 215.
Plant Protection, 108, 112, 115, 116, 117, 119, 125, 126, 163, 166, 168, 170, 171, 182, 185, 192.
Precautionary Approach, 73, 74, 109, 123.
Prevention, 79, 109, 110, 113, 114, 123, 134, 216.
Propagation, 67, 71, 161, 167, 171, 173, 183.
Protein, 73, 74, 140, 141, 152, 195, 218.

Q

Quercus Prinus, 32.
Quercus Velutina, 27.

R

Rainfall, 59, 70, 160, 173, 175, 181, 183, 186, 190, 206.
Recognition Species, 7, 8.
Research, 10, 75, 76, 94, 97, 109, 126, 128, 129, 136, 137, 146, 148, 155, 202, 216.
Robinia Pseudoacacia, 26, 50, 51, 67.
Roots, 24, 29, 31, 42, 53, 71, 72, 144, 145, 146, 149, 154, 156, 175, 176, 217.

S

Scales, 25, 29, 34, 41, 43, 46, 48, 84, 97, 171, 214, 219.
Seeds, 14, 24, 26, 27, 31, 33, 36, 38, 41, 44, 45, 46,

56, 60, 107, 113, 115, 118, 147, 161, 162, 163, 164, 165, 170, 171, 172, 180, 181, 182, 183, 186, 187, 188, 209, 212, 215, 217, 218, 219, 220, 221, 222, 224, 225, 226, 229.
Soil, 29, 38, 46, 55, 64, 65, 66, 71, 83, 84, 85, 112, 137, 138, 139, 140, 142, 146, 149, 150, 151, 152, 153, 154, 155, 156, 160, 161, 164, 165, 167, 168, 169, 171, 172, 173, 174, 175, 176, 181, 183, 186, 187, 188, 189, 190, 191, 203, 204, 213, 215, 219, 222, 227.
Stem, 23, 31, 37, 44, 57, 58, 60, 61, 95, 142, 168, 180, 185, 186, 217, 218, 224.
Strategies, 109, 115, 116, 123, 133, 206.
Striped Maple, 43, 44.
Structure, 176, 218, 219.
Sugar Maple, 27, 44, 49.
Symptoms, 138, 155, 192.

T

Tamarind, 80, 170, 171.
Temperature, 79, 85, 102, 153, 167, 169, 177, 178, 183, 186, 190.
Terminology, 134, 193.
Tibetan Medicines, 135, 136, 137.
Trade, 74, 76, 78, 79, 108, 109, 111, 113, 114, 119, 127, 128, 129, 133, 149, 195, 197, 209.
Treatment, 76, 92, 138, 188.
Tree Species, 20, 50.
Tulip Tree, 46, 69, 70.
Typological Species, 6.

U

University, 94, 97, 157, 158, 160, 167, 169, 170, 216.

V

Vanilla, 80, 173, 174, 175, 176, 177, 178, 179, 180.
Variations, 6, 10, 17, 36, 104, 106.
Vegetation, 138, 141, 151, 177.
Vegetative Propagation, 67, 71, 171.
Vitamins, 140, 141, 150.

W

Watering, 161, 184, 222.
Weeds, 153, 156, 169, 190, 212.
White Oak, 47.
Wildlife, 8, 22, 27, 29, 36, 37, 48, 57, 59, 64, 65, 68, 69, 212.
World Trade, 109, 119.

❑❑❑